Engineering Materials

This series provides topical information on innovative, structural and functional materials and composites with applications in optical, electrical, mechanical, civil, aeronautical, medical, bio- and nano-engineering. The individual volumes are complete, comprehensive monographs covering the structure, properties, manufacturing process and applications of these materials. This multidisciplinary series is devoted to professionals, students and all those interested in the latest developments in the Materials Science field, that look for a carefully selected collection of high quality review articles on their respective field of expertise.

Indexed at Compendex (2021) and Scopus (2022)

Khubab Shaker · Asif Hafeez
Editors

Advanced Functional Polymers

Synthesis to Applications

Springer

Editors
Khubab Shaker
Department of Materials
School of Engineering & Technology
National Textile University
Faisalabad, Pakistan

Asif Hafeez
Department of Materials
School of Engineering & Technology
National Textile University
Faisalabad, Pakistan

ISSN 1612-1317 ISSN 1868-1212 (electronic)
Engineering Materials
ISBN 978-981-99-0786-1 ISBN 978-981-99-0787-8 (eBook)
https://doi.org/10.1007/978-981-99-0787-8

This Springer imprint is published by the registered company Springer Nature Singapore Pte Ltd.
The registered company address is: 152 Beach Road, #21-01/04 Gateway East, Singapore 189721, Singapore

Contents

Introduction to "Advanced Functional Polymers: Synthesis to Applications"

Khubab Shaker and Asif Hafeez

Abstract Advanced functional polymers represent the class of materials with unique properties and applications in a variety of fields. This book covers recent advances in the synthesis and applications of advanced functional polymers including smart polymers, shape memory polymers, self-healing polymers, biopolymers, etc. The book also highlights the applications of these functional polymers in advanced functional polymers in various fields including lamellar membranes, paints, coatings, composite implants, zero-G applications, etc. The characterisation and recycling techniques for these polymers and composites are also discussed in the book.

Polymers-based materials are currently leading most of the current technological developments and these materials hold huge significance in many industries and applications. This book is targeted at advanced functional polymers, from synthesis to applications. The book has been split into ten different chapters covering the basic polymers to advanced application areas.

Chapter 2 covers the stereospecific polymerization techniques of different types of polymers, providing a brief overview related to the stereochemistry of polymers with mechanical, thermal, and electrical properties.

Chapter 3 is focused on Lamellar Membranes, providing a comprehensive review of two-dimensional (2D) nanomaterials that provide tremendous potential for separation applications. In lamellar membranes, mass transport occurs through the inherent interplanar spacing of the nanomaterials and the nanochannels generated during the fabrication of lamellar membranes. Various routes for preparing lamellar membranes including 2D nanosheet synthesis strategies, assembling techniques, mechanisms involved in transportation, and regulation of nanochannels are discussed in this chapter.

K. Shaker (✉) · A. Hafeez
Department of Materials, School of Engineering and Technology, National Textile University, Faisalabad, Pakistan
e-mail: khubab@ntu.edu.pk

A. Hafeez
e-mail: ahafeez@ntu.edu.pk

K. Shaker and A. Hafeez (eds.), *Advanced Functional Polymers*, Engineering Materials,
https://doi.org/10.1007/978-981-99-0787-8_1

Polymeric paints and coatings are addressed in Chap. 4. Polymeric coatings are applied on surfaces of metal, wood, plastics, and other products to offer protection, decoration, and special functions. The key trends in polymeric coatings, modern synthesis technologies, and other approaches are discussed in this chapter.

Chapter 5 covers smart polymers and composites used for various applications like fuel cells, bio-medical applications, temperature & pH-responsive polymers, membranes, and nanocomposites. The increased popularity of smart polymers and composites is attributed in terms of environmental friendliness, stimuli-responsive, and energy efficiency as a fuel cell and smart filtration process as a membrane.

Chapter 6 describes the self-healing elastomers, their complex structures, and various healing strategies. Elastomers with good mechanical performance and healing efficiency typically do not exist, as a self-healing mechanism limits the mechanical properties. This chapter focuses on self-healing elastomers in which covalent bonds in combination with noncovalent interactions provide the best balance between repairability and mechanical performance.

Chapter 7 presents biodegradable plastics of synthetic origin including aliphatic polyesters, aromatic copolymers, vinyl polymers, and biodegradable polyurethanes. Biodegradation (both biotic and abiotic) in synthetic-origin biodegradable polymers has been discussed followed by the classification and degradation tendencies in synthetic-origin biodegradable plastics.

Chapter 8 focuses on bioinspired polymer composite implants. Flexible fiber-reinforced composite prostheses respond to biologically friendly healing (secondary healing) and promote callus generation and soft tissue maturation and can provide solutions to problems. The performance of these composites from the published work according to different materials is discussed.

Functional polymers and composites for zero-gravity applications are deliberated in Chap. 9, including environmental conditions in zero-gravity (zero-G), Various structures used, and the material requirements for these structures. The commonly used materials for zero-gravity including polymers for thermal blankets, adhesives, polymer aerogels, shape memory polymers, fiber-reinforced composite materials, fiber metal laminates, protective coatings against atomic oxygen exposure, and lunar dust adhesion, etc. have been detailed in this chapter.

Chapter 10 focuses on the modern characterization techniques for functional polymers. Some of the advanced characterizations techniques discussed include Dynamic Mechanical Analysis (DMA), Thermal Mechanical Analysis (TMA), Atomic Force Microscopy (AFM), 4-Probe technique, Inverse Gas Chromatography (IGC), and Gel Permeation Chromatography (GPC).

The last chapter of the book targets the polymer and composite recycling techniques. Primary, secondary, and tertiary recycling techniques for thermoplastics are discussed in this chapter. Thermal, mechanical, and chemical waste recycling techniques for thermosets and the inherent recyclability of thermoset polymers are focused on in this chapter. The potential applications of recycled thermoplastics and thermosets are also discussed.

Stereospecific Polymerization Techniques

Asra Tariq and H. M. Fayzan Shakir

Abstract This chapter covers the stereospecific polymerization of different types of polymers that can be stereo regulated. The monomers of stereospecific polymers have side groups attached to the backbone of the polymer and their preparation involves complex catalytic systems. The characteristics of these polymers holding stereochemical isomerism in their main chain are mainly affected by the stereochemical structure. Up to now the control of stereochemistry to achieve the required properties is not fully realized. This chapter provides a brief overview related to the stereochemistry of polymers with mechanical, thermal, and electrical properties.

Keywords Stereoregularity · Polymerization · Vinyl polymers · Tacticity · Stereospecific polymers

1 Introduction

A stereospecific polymerization is a polymerization technique that controls the tacticity of the polymer formed. A stereo repeating unit is a configurational repeating unit containing a well-defined configuration at all sites of stereoisomerism. A stereoregular macromolecule builds a polymer that is an ordered macromolecule containing only one species of the stereo-repeating unit. A stereoregular polymer is fundamentally a tactic polymer. However, a tactic polymer is not always stereoregular, as a tactic polymer does not have defined sites of stereoisomerism [1–3].

During the last 50 years, various approaches of polymer synthesis were introduced where the configuration was either asymmetric or symmetric. More commonly, these polymers were synthesized using a catalyst that helps in configuration-specific polymer chain growth [4]. The stereospecific polymerization was first introduced in the early 1950s by Ziegler and Natta by introduction of innovative stereo-regulating polymerization systems for olefins [5, 6]. The tacticity of vinyl polymers was first

A. Tariq (✉) · H. M. Fayzan Shakir
Department of Materials, School of Engineering and Technology, National Textile University, Faisalabad, Pakistan
e-mail: asratariq23@gmail.com

K. Shaker and A. Hafeez (eds.), *Advanced Functional Polymers*, Engineering Materials,
https://doi.org/10.1007/978-981-99-0787-8_2

studied by proton Nuclear Magnetic Resonance (H NMR) for polymethyl methacrylate (PMMA) in 1960s [7], and is still considered a conclusive measure of the stereochemical polymeric configuration. Hence, this field of stereospecific polymerization developed alongside the growth of NMR spectroscopy technique [8].

2 Conventional Polymerization Techniques

The polymerization reaction can take place in all three states of matter (solid, liquid, gas). Usually, the solid-state reactions occur slowly and therefore are not commercially favorable. Most of the monomers existing in the gas phase can be polymerized by gas-phase reactions. However, the control of reaction parameters inside the reaction column is difficult due to complexity of the reaction. Owing to these limitations, liquid-phase reactions are preferred commercially. These reactions are generally divided as homogenous in-phase or heterogeneous phase reactions. The polymerization process occurring in a homogeneous phase is either bulk or solution polymerization, while polymerization in the heterogeneous phase can either be emulsion or suspension polymerization. The homogeneity or heterogeneity in polymerization reaction, implies the initial condition of the reaction system and is not a requirement throughout the reaction. A homogenous system before polymerization may become heterogeneous at the end of polymerization.

2.1 Bulk Polymerization

Bulk polymerization (mass polymerization) is conducted by adding a soluble initiator to a pure monomer in a molten state. Heat is provided to initiate the polymerization reaction. Figure 1 shows the schematic of bulk polymerization. During bulk polymerization, the viscosity increases over time due to increase in the molecular weight of the propagating polymer. This propagation of the reaction is highly exothermic and continuous stirring is required to avoid the buildup of excessive heat in the reactor. Stirring becomes difficult as the polymerization proceeds because of an excessive increase in viscosity. This may cause burning of polymers and explosion of the reactor. The bulk polymerization involves pure monomer without any solvent, hence it provides a high-purity polymer. This technique is used in many chain growth and step growth polymerization processes. Polymers prepared through this polymerization technique are PMMA and polystyrene [9, 10].

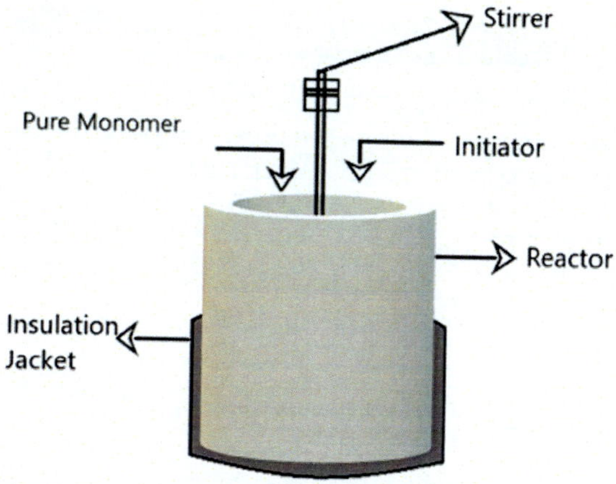

Fig. 1 Bulk polymerization process

2.2 *Solution Polymerization*

Mostly commercial polymerization is done by the solution polymerization process. In this technique, the monomer is dissolved in a solvent, already containing the catalyst. The produced polymer is dissolved in the solvent and removed at the end of the reaction. The presence of the solvent in the reactor controls the increase in viscosity, and heat is easily dissipated. The excess solvent is removed at the end of the reaction, and the pure polymer is obtained. The prepared polymer can still hold the solvent and is therefore preferred for applications like paints, varnishes, and adhesives. This polymerization technique is used for the production of polyacrylic acid (PAA) and polyacrylonitrile (PAN). The solvent selected for this polymerization should neither react with the monomer nor hinder the polymerization process [11]. Figure 2 exhibits the process of solution polymerization.

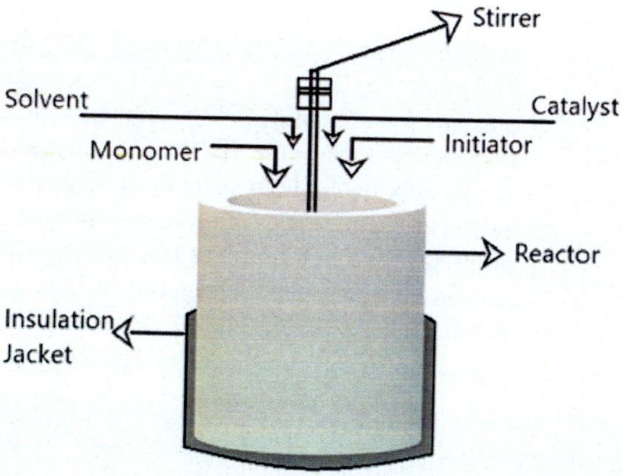

Fig. 2 Solution polymerization process

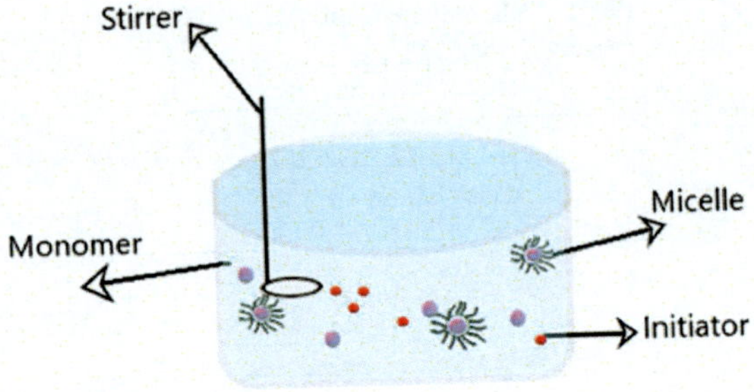

Fig. 3 Schematic of emulsion polymerization

2.3 *Emulsion Polymerization*

In emulsion polymerization, radical polymerization initiates in an emulsion phase having water, monomer, and surfactant. Usually, the emulsion is prepared by adding monomer (oil) to the water as the droplets, and is sustained using a surfactant. Some water soluble polymers such as polyvinyl alcohols (PVA) or hydroxyethyl cellulose, can be utilized as emulsifiers or stabilizers. The emulsion polymerization proceeds in the particles that develop spontaneously in the initial state of the process. The size of these particles is usually 100 nm and it holds many distinct polymer chains. Each particle is surrounded by a surfactant that resists the particles from coagulating with each other. Many commercial polymers are prepared by emulsion polymerization and the polymers used as solids are separated from the aqueous dispersion after polymerization. The polymers prepared through this polymerization are used in paints, adhesives, paper, and coatings. Figure 3 represents the emulsion polymerization process [12].

2.4 *Suspension Polymerization*

Suspension polymerization proceeds only through a radical polymerization reaction. The monomer being insoluble in water, remains in a suspended form. The Monomer appears as an organic medium and water as a continuous medium. The initiator such as a benzoyl peroxide (BOP), Azobisisobutyronitrile (AIBN), or $(NH_4)_2(S_xO_4)_y$ can be either soluble in water or monomer. Continuous agitation during the reaction assists in maintaining a two-phase system, by keeping the droplets apart. The reaction proceeds in a small droplet suspended in water just like the bulk polymerization and heat is released in water. This polymerization technique is used for the polymerization of butadiene, isoprene, isobutylene, vinyl chloride, and methyl methacrylate [13]. Figure 4 displays the suspension polymerization process.

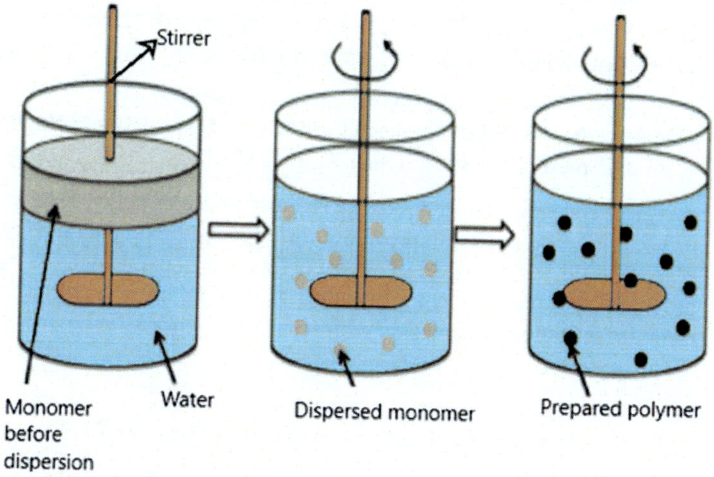

Fig. 4 Suspension polymerization process

3 Advanced Polymerization

Usually, commercial polymers are polymerized through these above-mentioned polymerization techniques by following chain-growth polymerization that includes free radical, anionic, or cationic polymerization or step growth polymerization reactions. These polymerization techniques offer most commodity polymers having broad molecular weight distribution with versatile properties. In conventional radical polymerization, polymer chains were continuously forming through a process called propagation and terminations started to occur. In the initial state the molecular weight was much higher and the rate of propagation was high, and with the depletion of monomer, the molecular weight of the prepared chains started to reduce. So, the polymer chains produced at the start of the reaction were higher in length than the ones formed in a later stage. Hence a broader molecular weight distribution is obtained through conventional radical polymerization. In ideal living radical polymerization, a specific number of monomers are activated once and these activated monomers propagate and chains start to form. Termination is done by a radical-to-radical combination process. However, in reality, all chains cannot be activated at the same time.

However, for some specific applications, a controlled polymerization reaction is required to optimize specific properties of polymers such as the architecture of polymers, the molecular weight of polymers, molecular weight distribution, and functionality of polymers. Through this controlled polymerization reaction, many of the reaction anomalies can be avoided such as premature termination, side chain reactions, or excessive radical formation. For this controlled polymerization process living radical polymerization (LRP), also called controlled radical polymerization

(CRP) was introduced. This controlled radical polymerization was further divided into the following fundamental synthesis process.

- Atom Transfer Radical Polymerization (ATRP)
- Reversible Addition/Fragmentation Chain Transfer Polymerization (RAFT)
- Amine–peroxide redox polymerization (APRP)

The polymer produced through ATRP or surface modification done by this polymerization process offers improvement in the properties of polymers because of adhesive properties, tailored hydrophilicity, or functionalization of the nanoparticle. These polymers prepared through either RAFT or ATRP also enable improvements for bio-mineralization, drug delivery, bio-compatibilization, and hydrogel applications.

3.1 Atom Transfer Radical Polymerization (ATRP)

Today's most successful polymer synthesis technique is atom transfer radical polymerization (ATRP) for the preparation of new polymers having defined and or narrow molecular weights, and narrow molecular weight distributions, with improved chain end functionalities. ATRP not only provided unprecedented control over molecular architecture but also offers the route to prepare polymers with complex architecture such as star-shaped polymers. ATRP is based on a defined catalytic system generally metal-based complex catalyst systems having transition metals [14–17]. Monomers that are polymerized through ATRP are vinyl chloride, styrene, acrylonitrile, and methyl methacrylate [18]. In the ATRP process, components required are monomers, initiators that can be halo-compounds as alkyl- or aryl- or AIBN, nitrogen, or phosphorus-based ligands, transition-metal salts, and solvents. The kinetics of the reaction is dependent on all these parameters [19]. The metal complex having a lower oxidation state reacts with the initiator (alkyl halide) and yields a radical R• along with a higher oxidation state metal complex, this step is called activation, The produced radicals can further react with the monomer-generating polymer with a constant specific rate called the rate of propagation. The step that distinguishes the ATRP from the conventional polymerization process is the production of the halogen-terminated polymeric dormant state, which is reactivatable for reaction with the transition metal [15, 20].

3.2 Reversible Addition/Fragmentation Chain Transfer Polymerization (RAFT)

Reversible addition-fragmentation chain transfer polymerization (RAFT) is an innovative and new type of controlled radical polymerization that provides high control on the molecular weight and molecular weight distribution of a polymer. A polymer prepared with this reaction can have a well-defined architecture (block, star, comb, brush, and gradient copolymers) or predictable functionality. This outcome of this RAFT approaches the ideal living radical polymerization with high convenience to perform as it only requires the addition of a thiocarbonylthio compound in the conventional radical polymerization. Almost all vinyl monomers can be polymerized through RAFT to obtain a defined molecular weight and narrow molecular weight distribution.

The organic compound that contains a thiocarbonylthio group-based functional group is called the RAFT agent. Majorly the RAFT agents developed and used belong to one of the following types: xanthates, dithioesters, dithiocarbamates, trithiocarbonates, and dithiophosphonates. Usually, the ideal RAFT process occurs with the RAFT agent working as a transfer agent. The living character of the RAFT is possible because of the retention of the thiocarbonylthio groups that allow the production of controlled architecture copolymers and polymers with the required functionality. The most popular RAFT agent is dithioesters. The process starts just like the free-radical chain-growth polymerization initiated by some initiator for example dibenzoyl peroxide (BPO) or azoisobisbutyronitrile (AIBN). The prepared initiator will start a few monomers to form a chain. The chain radicals are trapped by the RAFT agent and make an intermediate radical. An original initiating radical species is produced by the fragmentation of the intermediate. Fragmentation of intermediates can also produce polymeric RAFT agents and an organic radical. Initially, the fragmentation will only proceed in a forward reaction, and the reverse reaction is energetically not favorable. The high degree of polymerization through RAFT is possible by increasing the rate of the addition step as compared to the propagation step. The small number of growing chains and short activation cycles will highly minimize the number of termination reactions per cycle. Thiocarbonyl groups used as the RAFT agent will trap most of the remaining radicals and as a consequence, the dead chains 'number is only in control of by the number of radicals added initially during the start of polymerization [21–24].

Researchers are nowadays working on the dual control of both molecular weight and stereoregularity and it was achievable by stereospecific living polymerization. No doubt the dual control over stereoregularity and the molecular weight is difficult but still possible to obtain. In the type of cationic RAFT polymerization, various kinds of low nucleophilic counter anions can be utilized owing to the quickly reversible interconversion of dormant and cationic propagating species. It was reported that both atactic and isotactic poly(isobutyl vinyl ether)s having predictable molecular weights were produced using different counteranions [25].

4 Vinyl Polymerization

The vinyl monomer that is CH_2=CHX known as a-substituted vinyl monomers or CH_2=CYX known as a-disubstituted vinyl (vinylidene) monomers can be polymerized in a variety of ways. A common feature in these types of vinyl polymerization is that the growing chain retains an active center throughout its growth. The main chain having a carbon-to-carbon long-chain holding substituent group(s) is called "pseudo-asymmetric" as such carbons do not hold the four distinct substituent groups required to be truly asymmetric disregarding the chain ends [5, 6, 26–29]. The different arrangement of these substituent groups along the main carbon chain results in different tacticity of the polymer. A regularly arranged substituent by the side of a polymer chain is called an isotactic stereo structure, which provides the arrangement of all the substituent groups on the same side of the zigzag plane main chain. This structure is commonly symbolized by a "rotated Fischer projection" that has the main carbon-to-carbon chain backbone symbolized by a horizontal line as shown in Fig. 5.

A different regular arrangement is in which the substituent groups appear in alternate positions from side to side and as result, the neighboring units are shown opposite in configuration, this is called a syndiotactic structure, as shown in Fig. 6.

Irregular placement of substituent groups along the carbon main chain is called a-tactic structure as shown in Fig. 7.

The smallest unit showing the configuration of the repeating units is known as a diad. Vinyl polymer demonstrates two types of diads which are classified as meso

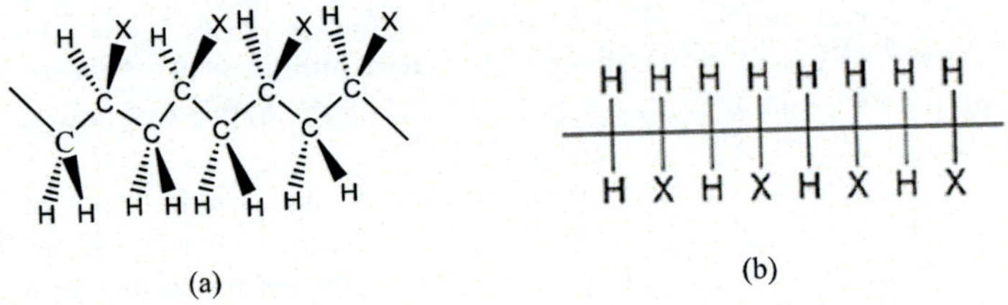

(a) (b)

Fig. 5 Isotactic vinyl polymer **a** zigzag chain representation **b** rotated Fischer expression

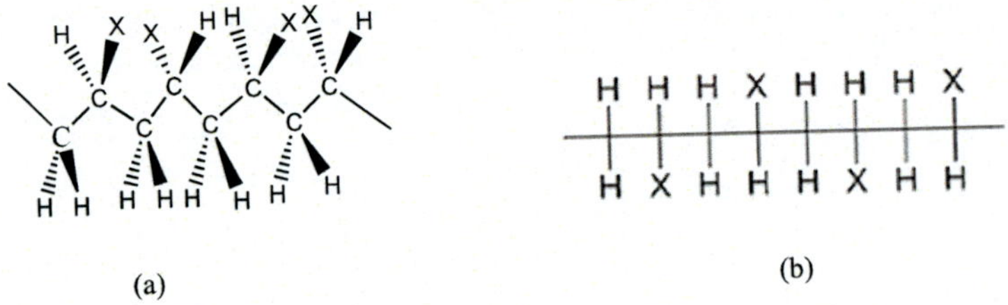

(a) (b)

Fig. 6 Syndiotactic vinyl polymer **a** zigzag chain representation **b** rotated Fischer demonstration

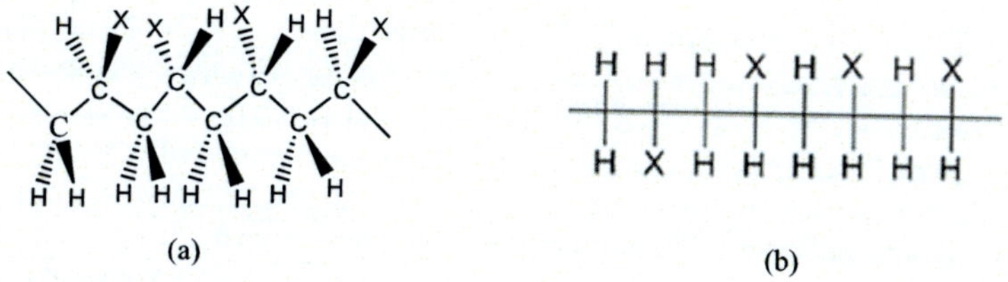

Fig. 7 A tactic vinyl polymer **a** zigzag chain representation **b** rotated Fischer demonstration

Fig. 8 Diad stereoisomerism **a** meso, m **b** racemo, r

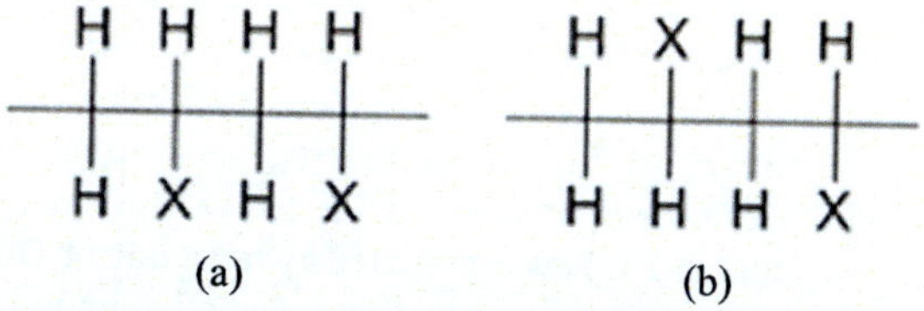

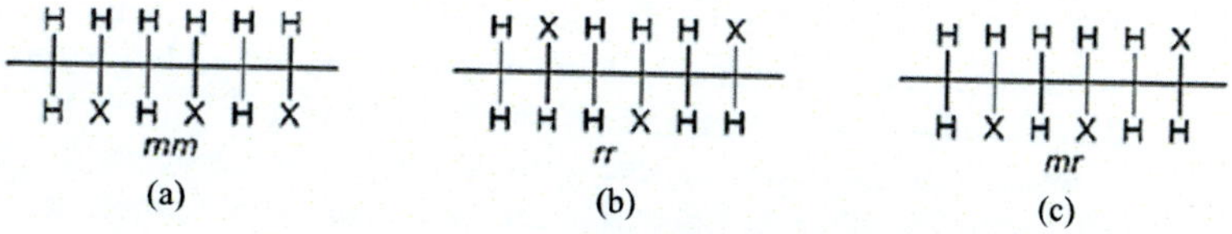

Fig. 9 Triad stereoisomerism, **a** Isotactic, **b** Syndiotactic, **c** Heterotactic

(m) and racemo (r) visible in Fig. 8. These notations are utilized as a sequence in an isotactic polymer and demonstrated as mmmmmm, however, in the syndiotactic polymer as –rrrrrr–. Practically, a polymer having a high content of isotactic or syndiotactic composition is hardly attainable, but the degree of regularity always needs to be studied [30].

The relative configurations of added monomers along the polymer chains can be defined by the notation of m and r, as a triad, as demonstrated in Fig. 9. Three possible triads, written by mm, rr, and mr formulate the polymer having isotactic, syndiotactic, and heterotactic triads, respectively. Heterotactic is a stereoregular type of polymer having heterotactic triads in the chain as shown in Fig. 9c.

In the same way, the six distinct structures are probable: mmm, mmr, rrr, rrm, rmr, and mrm [30].

4.1 *Stereospecific Isotactic Polymerization*

Monomer structure, initiator type, solvent, and reaction temperatures majorly play a role in the stereoregularity of polymers. As discussed earlier this specifically designed polymerization was introduced by Zeigler Nata for obtaining a regularly arranged side group such as isotactic polyolefins. A wide range of catalysts and catalytic systems were introduced for stereostructure control along with the development of heterogeneous Ziegler–Natta catalysis. One of the basic needs for the catalyst of stereospecific polymerization is the identification of monomers' probability to regulate the configuration of developing chiral carbons at growing chain ends. To make stereospecific isotactic polymers, rational catalyst designs have been established considerably in the homogeneous system usually polymerized by single-site organometallic catalysts, for example, metallocenes and related complex systems [31].

The most commonly employed catalysts for the production of isotactic polypropylene are synthesized by reacting titanium halide with alkylaluminum compounds. However, Ziegler developed a very advanced catalytic feature to control the stereospecificity of polypropylene by introducing a connection between the regularity of the surface of the heterogeneous catalyst and the polymer's regular structure. So, instead of synthesizing the catalyst by reacting TiCl with AlR_3, first, the crystalline $TiC1_3$ was prepared by reduction of TiCl with hydrogen with further treatment with trialkyl aluminum or a dialkyl aluminum chloride. This catalyst when used for preparing isotactic polypropylene, and increased the isotacticity in the product from almost 40–90% [31]. Treatment of the product obtained by reacting TiCI, and $AIRCl_2$ with Lewis bases (commonly ethers) was observed to be a larger surface area of the catalyst and with improved activity and high stereospecificity [32].

Isotactic polystyrene was developed by using a heterogeneous Ziegler–Natta catalysis system. Some researchers also explained the use of homogeneous systems that enabled the polymerization of styrene providing isotactic configuration catalyzed by the system as shown in Fig. 10. The application of anionic initiators holding a lithium ester enolate having high excess of lithium trimethylsilanolate (Me_3SiOLi) showed good control on the tacticity.

Polymerization of methacrylate was done extensively by containing the stereoregularity of the synthesized polymer. Researchers explained a variety of polymerization systems that can be utilized for the polymerization of methacrylates including anionic or also called living polymerization, radical, and organometallic catalysis systems. The polymerization of methacrylates by an anionic system with appropriate reaction parameters permits the preparation of various types of stereoregular polymethacrylate structures. The first initiator used for the isostatic polymerization of methacrylates was t-butylmagnesium bromide ($t\text{-}C_4H_9MgBr/MgBr_2$). This initiator offered a

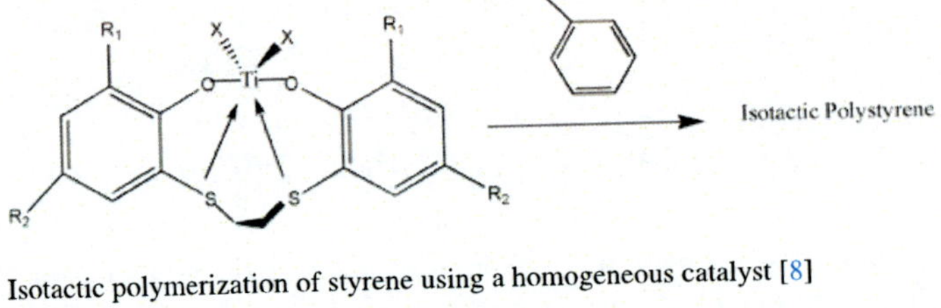

Fig. 10 Isotactic polymerization of styrene using a homogeneous catalyst [8]

considerably isotactic product (mm >95%) with a highly narrow molecular weight distribution (MWD) when polymerized by living polymerization [6].

Stereoregulation obtained through radical polymerization was enormously improved in recent years, such as the radical polymerization of isotactic N-isopropyl acrylamide by the use of lanthanide Lewis acids ($Lu(OTf)_3$, $Yb(OTf)_3$, and $Sc(OTf)_3$) [26]. A cationic initiator was added for the polymerization of vinyl ethers, however, few reports explained the high stereospecific cationic polymerization [27]. Tert-butyl vinyl ether when cationic polymerized using BF_3-based initiators in toluene solvent provided an isotactic polymer (mm =79%) that can be easily transformed to isotactic poly(vinyl alcohol).

4.2 *Stereospecific Syndiotactic Polymerization*

For the syndiotactic polymerization of polypropylene, catalytic systems explained up to now are the complexes specific for the isotactic polymerization of polypropylene. Some of these catalytic systems were obtained from the vanadium (V) and aluminum (Al) compounds. However, it was observed that the systems studied earlier are specifically for the stereospecific syndiotactic polymerization of propylene polymer only if the Al/V ratios are within a certain range. For instance, if the catalytic systems are designed based on aluminum dialkylmonochloride (or aluminum dialkylmonofluoride) and vanadium triacetylacetonate for stereospecific polymerization of propylene to a syndiotactic polymer at a ratio of Al/V approximately 5, the temperature should be below 0 °C [33].

Polymerization of syndiotactic polystyrene having a high percentage of syndiotacticity, (more than 94%) and a high degree of crystallinity is synthesized in crystalline form in a planar zigzag conformation [34]. Using methyl aluminoxane along with titanium mixture as a catalyst compound results in the formation of syndiotactic polystyrene even above the room temperature. The mechanism of this polymerization was proposed to be slightly different as compared to the other catalyst systems working at low temperatures and usually works by a radical pathway. The polymerization of syndiotactic polypropylene using Ziegler–Natta catalysts is proceeded usually at temperatures below −65 °C [33, 35, 36]. Titanium-based compounds such as $TiCl_4$, $Ti(OEt)_4$, or (η-C_5H_5)$TiCl_3$) along with methyl aluminoxane act as a catalyst for the polymerization of styrene, above room temperature and synthesize pure syndiotactic polystyrene having very narrow MWD. Ring-substituted styrenes are also used for the preparation of pure syndiotactic polymers. The reactivity of the monomer is improved by the electron-releasing substituents over the aromatic ring [37].

The polymerization of methyl methacrylate to prepare highly syndiotactic polymer by living anionic polymerization was done using the combinations t-C_4H_9Li and R_3Al in toluene solvent at −78 °C [38]. The catalyst t-C4H9Li when used alone offered a polymer majorly isotactic having broad MWD. Adding R_3Al, for example,

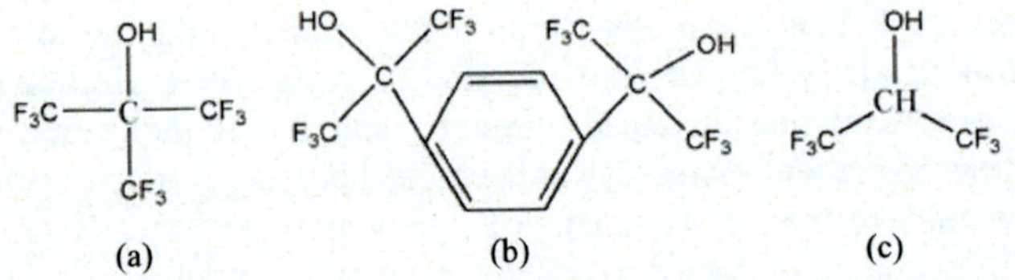

Fig. 11 Fluorinated alcohols are utilized for the preparation of stereospecific polymer by radical polymerization [8]

$(n\text{-}C_4H_9)_3Al$ with the ratio of Al/Li ≥ 3, the polymer prepared resulted in 90% syndiotacticities and the number average molecular weight (M_n) value was almost the same as obtained using $t\text{-}C_4H_9Li$. A bulky methacrylate monomer such as trimethylsilyl methacrylate (TMSMA) provided highly syndiotactic polymers by polymerizing with $t\text{-}C_4H_9Li$/bulky aluminum phenoxide. The synthesized polymer at 95 °C was 98% syndiotactic [39].

Researchers explained the use of some polar solvents as a good reaction medium to obtain controlled Stereoregulation in radical polymerization [40–42]. The solvents can interact with the polar side of the monomers and the chain propagates resulting in hydrogen bonding as the chain ends. From these polar solvents, fluorinated alcohol can be used as a strong hydrogen donor. Figure 11a shows the solvent used in the radical polymerization of vinyl acetate that offered syndiotactic polymer [r = 73% (−78 °C), r = 62% (20 °C)].

Figure 11b presents the solvents used for the syntheses of highly stereoregular PMMA (r = 88%, 20 °C) and Fig. 11c indicates the solvents for poly(- N-methylmethacrylate) (r = 98%, 78 °C) that were described. This stereospecificity in polymers is obtained due to the presence of electrostatic repulsions in the carbonyl polar groups and fluorinated alcohols hydrogen bonding the propagating chain end with the attachment of new polymers.

Researchers quoted that the polymerization of Poly(N-isopropyl acrylamide) by hydrogen-bond assisted stereospecific radical polymerizations using Lewis bases [43–50] and alcohol compounds [51, 52] helps in stereoregulation. Adding hexamethylphosphoramide in the polymerization of Poly(N-isopropyl acrylamide) in toluene solvent at −60 °C results in 72% racemo (r) [45].

The hydrogen acceptors and or Lewis basic solvents interact with the monomer molecules that hold hydrogen donating units and offer a stereoregular polymer [42].

4.3 Stereospecific Heterotactic Polymerization

Heterospecific polymerization shows high-order stereoregulation as compared to isotactic and syndiotactic stereospecific polymer since in this type of polymer the **m** and **r** come alternately to give regularity in structure. A very limited study was done on heterospecific polymerization by anionic or radical polymerization. Heterotactic-specific polymerization by living anionic polymerization was stated for primary alkyl methacrylates using a catalyst system of t-C_4H_9Li in toluene solvent at low temperatures. The resulting polymer offered ***mr*** triad content up to 95% using controlled reaction parameters [39].

The polymerization for the preparation of polyvinyl alcohol (PVA) and polyvinyl acetate by cyclopolymerization using vinyl trialkylsilyl ethers and vinyl trialkylcarbinyl ethers for the synthesis of 60% heterotactic polymers was studied. First poly(viny1 trialkylsilyl ether) was prepared and then later on converted to PVA by treating it with hydrofluoric acid [53]. Heterotactic polymerization of N-n-propylmethacrylamide was obtained by the radical polymerizing in the presence of alkali metal bis(trifluoromethanesulfonyl) imides ($MNTf_2$ and toluene solvent affected the stereospecificities using $LiNTf_2$ [54]. Heterospecific anionic polymerization of N, N-dimethyl acrylamide was done by using 1,1-bis (40-trimethylsilylphenyl)-3,3-diphenyl propyl potassium/Et_2Zn at 18 °C in tetrahydrofuran solvent. Et_3B was also used instead of Et_2Zn which offered a better heterotactic polymer. Fluorinated alcohol-based solvents for the preparation of heterospecific radical polymerization were reported to be affected [26].

A Heterospecific polymer of vinyl pivalate in the presence of (n-C=4H_9)$_3$B with an air environment using $(CF_3)_3COH$ at 40 °C was reported with mr = 61.0%. Toluene and fluorinated alcohol solvents when used for the polymerization of N-isopropyl acrylamide offered heterospecific polymer at a temperature of 40 °C with mr = 75% [8].

5 Stereoregularity and Polymer Properties

Along with the explanation of stereoregularity and stereochemistry of advanced polymers, there is also a need to understand the relationship between the stereospecificity of polymers and the properties of polymers. Synthesis of new materials with controlled stereochemistry will result in the innovation of novel materials with customized behavior and function. The major bulk properties of polymers affected by controlling the stereoregularity of polymers are discussed below.

5.1 *Thermomechanical Properties*

Crystallinity is the most important factor which contributes to the thermomechanical properties of polymer materials. Crystallinity is the regularity or arrangement in the polymer structure at the molecular or chain level which highly depends on the tacticity. The properties of polypropylene are largely dependent on the tacticity of the polymer. Isotactic PP is hard and excellent in mechanical strength because good packing of chains result in high % crystallinity, however, atactic PP is soft, rubbery, high in ductility, and amorphous in nature. The thermal properties for example melting temperature (Tm) and glass transition temperature (Tg) also vary considerably because of the tacticity and crystallinity of the polymer [55]. The Tm of isotactic PP is around 160 °C because of ordered chain packing. However, because these two different isomers offer highly different mechanical and thermal properties, usually neither of them offers the best properties. Generally, a combination of both is developed having some degree of stereoregularity and stereoregularity Natta. In this type of PP, both isotactic and atactic structure offers high tensile strength and elongation and are called thermoplastic elastomers [56–62]. The syndiotactic PP developed for better crystallinity than atactic, but this material isn't strong mechanically as isotactic PP. The combination of the amorphous phase with the crystalline phase also helped in the adjustment of the processing parameters of PP as viscoelasticity has a direct influence on the molecular structure of PP chains [63].

In terms of industrial use, PS is a well-known commodity plastic. Only the amorphous, atactic form of PS was known until 1960 and then Nata introduced Isotactic PS [64]. This isomer of PS shows higher mechanical properties when compared to the former atactic because of its ability to get crystallized (60% crystallinity, Tm ≈250 °C) if it is annealed and this difference in mechanical properties becomes even more significant if compared to temperatures greater than their glass transition temperatures. The other thermal properties like glass transition temperature (T_g) and heat capacity of both types of PS are almost similar. However, isotactic PS has not been acknowledged in commercial plastic due to it lower rate of crystallization [65].

In 1986, syndiotactic PS was discovered and it showed significant differences in characteristics when compared to its previous isomers. Syndiotactic PS has a high melting temperature as compared to isotactic PS (Nearly 50 °C higher) [66]. The rate of the crystallization process is high for syndiotactic PS as compared to isotactic PS [67, 68]. Syndiotactic PS and isotactic PS are thermally more stable in comparison to atactic isomers, showing that stereochemistry affects the thermal degradation profiles [69]. In addition to the high rate of the crystallization process, syndiotactic PS also possesses the appealing attributes of atactic PS, which are good moldability, improved electrical properties, and resistance to hydrolytic degradation [70]. The entangled molecular chains eventually cause more robust material because of the phenyl groups' higher rotational energy barriers [71]. One of the copolymers was introduced as a stereoblock of syndiotactic and atactic in which the syndiotactic content controlled the thermal properties such as T_m and crystallization rates [72, 73].

The thermal properties of PMMA varied drastically by changing the stereochemistry between syndiotactic and isotactic analogs such that the difference in the T_g of these polymers was greater than 70 °C [74–76]. Isotactic PMMA has T_g much lower than syndiotactic PMMA, because the former is high in crystallinity, while the latter is amorphous in structure, and its properties resemble atactic PMMA [77, 78].

5.2 *Electrical Properties*

The electrical properties of many conductive polymers also vary by changing the stereoregularity of the polymer. Polyacetylene is a famous conductive polymer having a chiral side chain [79]. However, attaching large side chains to polyacetylene tends to affect the conjugative pathway along the polymer main chain, which renders it inappropriate for organic semiconductor applications. Hence the polymers with conjugated backbone were introduced which maintain their electrical conductivity even when changing the attached substituent groups such as polypyrrole, polythiophenes, and polyanilines [80]. These conductive polymers showed great potential for sensing applications [81, 82]. The most apparent advantage offered by acrylic and other vinyl polymers is their wide potential derivatives due to the ease of functionalization and reliable polymerization chemistry [55].

6 Conclusion

The stereospecific polymerization of different types of polymers using conventional, advanced, and vinyl polymerization was discussed in the chapter. Some of these techniques involve the use of monomer in the molten state or gaseous state. However the commercially viable option is the liquid phase where the polymer reaction is initiated by using some appropriate initiator. The chains are propagated which increases the viscosity and heat generation in the reactor. The heat dissipation can be facilitated by using monomer in solvents or while compromising the purity of the polymer. These polymers generally find the application in paints, varnishes and adhesives. The characteristics of polymers holding stereochemical isomerism in their main chain are mainly affected by the stereochemical structure. Many bulk properties of stereospecific polymers are interconnected with the stereoregularity of polymers such as % crystallinity, mechanical stability, thermal stability gas permeation, and electrical conductivity.

References

1. IUPAC: Basic definitions of terms relating to polymers. Pure Appl. Chem. **40**, 477 (1974)
2. IUPAC: Commission on macromolecular nomenclature. Pure Appl. Chem. **51**, 1101 (1979)
3. Penczek, S., Moad, G.: Glossary of terms related to kinetics, thermodynamics, and mechanisms of polymerization (IUPAC Recommendations 2008). Pure Appl. Chem. **80**(10), 2163–2193 (2008)
4. Ketley, A., Werber, F.: Stereospecific polymerization: a revolution in polymer synthesis has occurred in the last decade. Science **145**(3633), 667–673 (1964)
5. Catalysts, Z., Fink, G., Brintzinger, H.: Springer, Berlin, 1995 (2003); (b) Mülhaupt, R.: Macromol. Chem. Phys. 204:289–327

6. Matyjaszewski, K., Braunecker, W.A.: Radical polymerization. Macromolecular Engineering: Precise Synthesis, Materials Properties, Applications, pp. 161–215 (2007)
7. Ibbett, R.N.: NMR Spectroscopy of Polymers. Springer Science & Business Media (2012)
8. Hirano, T., Kitaura, T.: Stereospecific polymerization. In: Kobayashi, S., Müllen, K. (eds.) Encyclopedia of Polymeric Nanomaterials. Springer Berlin Heidelberg, Berlin, Heidelberg, pp. 2282–2303 (2015). https://doi.org/10.1007/978-3-642-29648-2_216
9. Cheremisinoff, N.P.: B. In: Cheremisinoff, N.P. (ed.) Condensed Encyclopedia of Polymer Engineering Terms. Butterworth-Heinemann, Boston, pp. 32–38 (2001). https://doi.org/10.1016/B978-0-08-050282-3.50007-X
10. Bhat, G., Kandagor, V.: 1—Synthetic polymer fibers and their processing requirements. In: Zhang, D. (ed.) Advances in Filament Yarn Spinning of Textiles and Polymers. Woodhead Publishing, pp. 3–30 (2014). https://doi.org/10.1533/9780857099174.1.3
11. Wan Ibrahim, W.H.B., Mujtaba, I.M.: Dynamic optimization of solution polymerization process of methyl methacrylate in batch reactors. In: Karimi, I.A., Srinivasan, R. (eds.) Computer Aided Chemical Engineering, vol. 31. Elsevier, pp. 1326–1330 (2012). https://doi.org/10.1016/B978-0-444-59506-5.50096-1
12. Lovell, P.A., Schork, F.J.: Fundamentals of emulsion polymerization. Biomacromol **21**(11), 4396–4441 (2020). https://doi.org/10.1021/acs.biomac.0c00769
13. Chaudhary, V., Sharma, S.: Suspension polymerization technique: parameters affecting polymer properties and application in oxidation reactions. J. Polym. Res. **26**(5), 102 (2019). https://doi.org/10.1007/s10965-019-1767-8
14. Matyjaszewski, K., Tsarevsky, N.V.: Nanostructured functional materials prepared by atom transfer radical polymerization. Nat. Chem. **1**(4), 276–288 (2009)
15. Tsarevsky, N.V., Matyjaszewski, K.: "Green" atom transfer radical polymerization: from process design to preparation of well-defined environmentally friendly polymeric materials. Chem. Rev. **107**(6), 2270–2299 (2007)
16. Matyjaszewski, K., Xia, J.: Atom transfer radical polymerization. Chem. Rev. **101**(9), 2921–2990 (2001)
17. Wang, J.-S., Matyjaszewski, K.: Controlled/"living" radical polymerization. Atom transfer radical polymerization in the presence of transition-metal complexes. J. Am. Chem. Soc. **117**(20), 5614–5615 (1995)
18. Chmielarz, P., Król, P.: Chapter 10—Poly(urethane-methacrylate) copolymers prepared by the atom transfer radical polymerization methods as a new material for hydrophobic coatings. In: Thomas, S., Datta, J., Haponiuk, J.T., Reghunadhan, A. (eds.) Polyurethane Polymers. Elsevier, Amsterdam, pp. 247–260 (2017). https://doi.org/10.1016/B978-0-12-804039-3.00010-5
19. Camacho-Cruz, L.A., Velazco-Medel, M.A., Bucio. E.: Chapter 9—Aqueous polymerizations. In: Inamuddin, Boddula, R., Asiri, A.M. (eds.) Green Sustainable Process for Chemical and Environmental Engineering and Science. Elsevier, pp. 275–318 (2020). https://doi.org/10.1016/B978-0-12-819542-0.00009-9
20. Tsarevsky, N.V., Tang, W., Brooks, S.J., Matyjaszewski, K.: Factors determining the performance of copper-based atom transfer radical polymerization catalysts and criteria for rational catalyst selection. ACS Publications (2006)
21. Chiefari, J., Chong, Y., Ercole, F., Krstina, J., Jeffery, J., Le, T.P., Mayadunne, R.T., Meijs, G.F., Moad, C.L., Moad, G.: Living free-radical polymerization by reversible addition−fragmentation chain transfer: the RAFT process. Macromolecules **31**(16), 5559–5562 (1998)
22. Moad, G., Rizzardo, E., Thang, S.H.: Living radical polymerization by the RAFT process. Aust. J. Chem. **58**(6), 379–410 (2005)
23. Rizzardo, E., Moad, G., Thang, S.H.: Reversible addition–fragmentation chain transfer polymerization. In: Encyclopedia of Polymer Science and Technology (2002)
24. Keddie, D.J.: A guide to the synthesis of block copolymers using reversible-addition fragmentation chain transfer (RAFT) polymerization. Chem. Soc. Rev. **43**(2), 496–505 (2014)
25. Uchiyama, M., Satoh, K., Kamigaito, M.: Stereospecific cationic RAFT polymerization of bulky vinyl ethers and stereoblock poly(vinyl alcohol) via mechanistic transformation to radical

RAFT polymerization of vinyl acetate. Giant **5**, 100047 (2021). https://doi.org/10.1016/j.giant.2021.100047
26. Matyjaszewski, K.: Atom transfer radical polymerization (ATRP): current status and future perspectives. Macromolecules **45**(10), 4015–4039 (2012)
27. Kanazawa, A., Kanaoka, S., Aoshima, S.: A stepping stone to stereospecific living cationic polymerization: cationic polymerization of vinyl ethers using iron (II) sulfate. J. Polym. Sci. Part A Polym. Chem. **48**(16), 3702–3708 (2010)
28. Razavi, A., Thewalt, U.: Site selective ligand modification and tactic variation in polypropylene chains produced with metallocene catalysts. Coord. Chem. Rev. **250**(1–2), 155–169 (2006)
29. Song, E.-H., Shang, J., Ratner, D.: Polysaccharides (2012)
30. Hirano, T., Kitaura, T.: Stereospecific polymerization. In: Kobayashi, S., Müllen, K. (eds.) Encyclopedia of polymeric nanomaterials. Springer Berlin Heidelberg, Berlin, Heidelberg, pp. 1–25 (2021). https://doi.org/10.1007/978-3-642-36199-9_216-1
31. Pino, P., Mülhaupt, R.: Stereospecific polymerization of propylene: an outlook 25 years after its discovery. Angew. Chem. Int. Ed. Engl. **19**(11), 857–875 (1980)
32. Quirk, R.P.: Transition Metal Catalyzed Polymerizations: Alkenes and Dienes: Papers Presented at the Eleventh Midland Macromolecular Meeting Held August 17–21, 1981 in Midland, Michigan; Edited by Roderic P. Quirk ... [et al.]. MMI Press (1983)
33. Natta, G., Pasquon, I., Zambelli, A.: Stereospecific catalysts for the head-to-tail polymerization of propylene to a crystalline syndiotactic polymer. J. Am. Chem. Soc. **84**(8), 1488–1490 (1962). https://doi.org/10.1021/ja00867a029
34. Ishihara, N., Seimiya, T., Kuramoto, M., Uoi, M.: Crystalline syndiotactic polystyrene. Macromolecules **19**(9), 2464–2465 (1986)
35. Doi, Y., Ueki, S., Keii, T.: "Living" coordination polymerization of propene initiated by the soluble V (acac) 3-Al (C2H5) 2Cl system. Macromolecules **12**(5), 814–819 (1979)
36. Doi, Y., Suzuki, S., Soga, K.: Living coordination polymerization of propene with a highly active vanadium-based catalyst. Macromolecules **19**(12), 2896–2900 (1986)
37. Ishihara, N., Kuramoto, M., Uoi, M.: Stereospecific polymerization of styrene giving the syndiotactic polymer. Macromolecules **21**(12), 3356–3360 (1988)
38. Matyjaszewski, K., Gnanou, Y., Leibler, L.: Macromolecular Engineering. Wiley Online Library (2011)
39. Kitayama, T., Horii, H., Hatada, K., Kashiwagi, T.: Tacticity dependence of thermal degradation of PMMAs with the same chemical structure. Polym. Bull. **21**(4), 433–437 (1989)
40. Satoh, K., Kamigaito, M.: Stereospecific living radical polymerization: dual control of chain length and tacticity for precision polymer synthesis. Chem. Rev. **109**(11), 5120–5156 (2009)
41. Kamigaito, M., Satoh, K., Wan, D., Sugiyama, Y., Koumura, K., Shibata, T., Okamoto, Y.: Stereospecific living radical polymerization. ACS Publications (2006)
42. Hirano, T., Kamikubo, T., Fujioka, Y., Sato, T.: Hydrogen-bond-assisted syndiotactic-specific radical polymerization of N-isopropylacrylamide: the solvent effect on the stereospecificity. Eur. Polymer J. **44**(4), 1053–1059 (2008). https://doi.org/10.1016/j.eurpolymj.2008.02.002
43. Hirano, T., Miki, H., Seno, M., Sato, T.: Significant effect of hydrogen-bonding interaction on syndiotactic-specificity in radical polymerization of N-isopropylacrylamide. J. Polym. Sci. Part A Polym. Chem. **42**(17), 4404–4408 (2004)
44. Hirano, T., Miki, H., Seno, M., Sato, T.: Direct synthesis of syndiotactic-rich poly (N-isopropylacrylamide) via radical polymerization of hydrogen-bond-complexed monomer. Polymer **46**(11), 3693–3699 (2005)
45. Hirano, T., Miki, H., Seno, M., Sato, T.: Effect of polymerization conditions on the syndiotactic-specificity in radical polymerization of N-isopropylacrylamide and fractionation of the obtained polymer according to the stereoregularity. Polymer **46**(15), 5501–5505 (2005)
46. Hirano, T., Ishii, S., Kitajima, H., Seno, M., Sato, T.: Hydrogen-bond-assisted stereocontrol in the radical polymerization of N-isopropylacrylamide with primary alkyl phosphate: the effect of the chain length of the straight ester group. J. Polym. Sci. Part A Polym. Chem. **43**(1), 50–62 (2005). https://doi.org/10.1002/pola.20475

47. Hirano, T., Kitajima, H., Seno, M., Sato, T.: Hydrogen-bond-assisted stereocontrol in the radical polymerization of N-isopropylacrylamide with bidentate Lewis base. Polymer **47**(2), 539–546 (2006)
48. Hirano, T., Ishizu, H., Seno, M., Sato, T.: Hydrogen-bond-assisted isotactic-specific radical polymerization of N-isopropylacrylamide with pyridine N-oxide. Polymer **46**(24), 10607–10610 (2005)
49. Hirano, T., Ishizu, H., Sato, T.: Metal-free isotactic-specific radical polymerization of N-isopropylacrylamide with pyridine N-oxide derivatives: the effect of methyl substituents of pyridine N-oxide on the isotactic specificity and the proposed mechanism for the isotactic-specific radical polymerization. Polymer **49**(2), 438–445 (2008)
50. Hirano, T., Ishizu, H., Yamaoka, R., Ute, K., Sato, T.: Metal-free isotactic-specific radical polymerization of N-alkylacrylamides with 3, 5-dimethylpyridine N-oxide: the effect of the N-substituent and solvent on the isotactic specificity. Polymer **50**(15), 3522–3527 (2009)
51. Hirano, T., Okumura, Y., Kitajima, H., Seno, M., Sato, T.: Dual roles of alkyl alcohols as syndiotactic-specificity inducers and accelerators in the radical polymerization of N-isopropylacrylamide and some properties of syndiotactic poly (N-isopropylacrylamide). J. Polym. Sci. Part A Polym. Chem. **44**(15), 4450–4460 (2006)
52. Hirano, T., Kamikubo, T., Okumura, Y., Sato, T.: Heterotactic poly (N-isopropylacrylamide) prepared via radical polymerization in the presence of fluorinated alcohols. Polymer **48**(17), 4921–4925 (2007)
53. Nozakura, S.I., Ishihara, S., Inaba, Y., Matsumura, K., Murahashi, S.: Heterotactic poly (vinyl alcohol). J. Polym. Sci. Polym. Chem. Ed. **11**(5), 1053–1067 (1973)
54. Hirano, T., Segata, T., Hashimoto, J., Miwa, Y., Oshimura, M., Ute, K.: Syndiotactic-and heterotactic-specific radical polymerization of Nn-propylmethacrylamide complexed with alkali metal ions. Polym. Chem. **6**(27), 4927–4939 (2015)
55. Worch, J.C., Prydderch, H., Jimaja, S., Bexis, P., Becker, M.L., Dove, A.P.: Stereochemical enhancement of polymer properties. Nat. Rev. Chem. **3**(9), 514–535 (2019). https://doi.org/10.1038/s41570-019-0117-z
56. Natta, G.: Properties of isotactic, atactic, and stereoblock homopolymers, random and block copolymers of α-olefins. J. Polym. Sci. **34**(127), 531–549 (1959)
57. Giller, C., Gururajan, G., Wei, J., Zhang, W., Hwang, W., Chase, D.B., Rabolt, J.F., Sita, L.R.: Synthesis, characterization, and electrospinning of architecturally-discrete isotactic–atactic–isotactic triblock stereoblock polypropene elastomers. Macromolecules **44**(3), 471–482 (2011)
58. Harney, M.B., Zhang, Y., Sita, L.R.: Discrete, multiblock isotactic–atactic stereoblock polypropene microstructures of differing block architectures through programmable stereomodulated living Ziegler-Natta polymerization. Angew. Chem. Int. Ed. **45**(15), 2400–2404 (2006)
59. Madkour, T.M., Mark, J.E.: Mesoscopic modeling of the polymerization, morphology, and crystallization of stereoblock and stereoregular polypropylenes. J. Polym. Sci. Part B Polym. Phys. **40**(9), 840–853 (2002)
60. Coates, G.W., Waymouth, R.M.: Oscillating stereocontrol: a strategy for the synthesis of thermoplastic elastomeric polypropylene. Science **267**(5195), 217–219 (1995)
61. Mallin, D.T., Rausch, M.D., Lin, Y.G., Dong, S., Chien, J.C.: rac-[Ethylidene (1-. eta. 5-tetramethylcyclopentadienyl)(1-. eta. 5-indenyl)] dichlorotitanium and its homopolymerization of propylene to crystalline-amorphous block thermoplastic elastomers. J. Am. Chem. Soc. **112**(5), 2030–2031 (1990)
62. Collette, J.W., Tullock, C.W., MacDonald, R.N., Buck, W.H., Su, A.C., Harrell, J.R., Mulhaupt, R., Anderson, B.C.: Elastomeric polypropylenes from alumina-supported tetraalkyl Group IVB catalysts. 1. Synthesis and properties of high molecular weight stereoblock homopolymers. Macromolecules **22**(10), 3851–3858 (1989)
63. Ahmad, N., Di Girolamo, R., Auriemma, F., De Rosa, C., Grizzuti, N.: Relations between stereoregularity and melt viscoelasticity of syndiotactic polypropylene. Macromolecules **46**(19), 7940–7946 (2013)

64. Natta, G., Corradini, P., Bassi, I.: Crystal structure of isotactic poly-alpha-butene. Nuovo Cimento **15**(1), 52–67 (1960)
65. Schellenberg, J.: Syndiotactic Polystyrene: Synthesis, Characterization, Processing, and Applications (2009)
66. Woo, E.M., Sun, Y.-S., Yang, C.-P.: Polymorphism, thermal behavior, and crystal stability in syndiotactic polystyrene vs. its miscible blends. Prog. Polym. Sci. **26**(6), 945–983 (2001)
67. Wang, C., Lin, C.-C., Tseng, L.-C.: Miscibility, crystallization and morphologies of syndiotactic polystyrene blends with isotactic polystyrene and with atactic polystyrene. Polymer **47**(1), 390–402 (2006)
68. Cimmino, S., Di Pace, E., Martuscelli, E., Silvestre, C.: Syndiotactic polystyrene: crystallization and melting behaviour. Polymer **32**(6), 1080–1083 (1991)
69. Chen, K., Harris, K., Vyazovkin, S.: Tacticity as a factor contributing to the thermal stability of polystyrene. Macromol. Chem. Phys. **208**(23), 2525–2532 (2007)
70. Ishihara, N., Kuramoto, M.: 32. Syntheses and properties of syndiotactic polystyrene. In: Studies in Surface Science and Catalysis, vol. 89. Elsevier, pp. 339–350 (1994)
71. Huang, C.-L., Chen, Y.-C., Hsiao, T.-J., Tsai, J.-C., Wang, C.: Effect of tacticity on viscoelastic properties of polystyrene. Macromolecules **44**(15), 6155–6161 (2011)
72. Annunziata, L., Monasse, B., Rizzo, P., Guerra, G., Duc, M., Carpentier, J.-F.: On the crystallization behavior of syndiotactic-b-atactic polystyrene stereodiblock copolymers, atactic/syndiotactic polystyrene blends, and aPS/sPS blends modified with sPS-b-aPS. Mater. Chem. Phys. **141**(2–3), 891–902 (2013)
73. Annunziata, L., Sarazin, Y., Duc, M., Carpentier, J.F.: Well-defined syndiotactic polystyrene-b-atactic polystyrene stereoblock polymers. Macromol. Rapid Commun. **32**(9–10), 751–757 (2011)
74. Ute, K., Miyatake, N., Hatada, K.: Glass transition temperature and melting temperature of uniform isotactic and syndiotactic poly (methyl methacrylate) s from 13mer to 50mer. Polymer **36**(7), 1415–1419 (1995)
75. Biroš, J., Larina, T., Trekoval, J., Pouchlý, J.: Dependence of the glass transition temperature of poly (methyl methacrylates) on their tacticity. Colloid Polym. Sci. **260**(1), 27–30 (1982)
76. Bywater, S., Toporowski, P.: Effect of stereostructure on glass transition temperatures of poly (methyl methacrylate). Polymer **13**(3), 94–96 (1972)
77. Gillham, J., Stadnicki, S., Hazony, Y.: Low-frequency thermomechanical spectrometry of polymeric materials: tactic poly (methyl methacrylates). J. Appl. Polym. Sci. **21**(2), 401–424 (1977)
78. Stadnicki, S., Gillham, J., Hazony, Y.: Low-Frequency Thermomechanical Spectrometry of Polymeric Materials: Tactic Polymethylmethacrylates. Princeton Univ NJ (1976)
79. Ciardelli, F., Benedetti, E., Pieroni, O.: Polymerization of racemic and optically active 4-methyl-1-hexyne. Die Makromolekulare Chemie: Macromol. Chem. Phys. **103**(1), 1–18 (1967)
80. Pu, L.: The study of chiral conjugated polymers. Acta Polym. **48**(4), 116–141 (1997)
81. Inal, S., Rivnay, J., Suiu, A.-O., Malliaras, G.G., McCulloch, I.: Conjugated polymers in bioelectronics. Acc. Chem. Res. **51**(6), 1368–1376 (2018)
82. Balint, R., Cassidy, N.J., Cartmell, S.H.: Conductive polymers: Towards a smart biomaterial for tissue engineering. Acta Biomater. **10**(6), 2341–2353 (2014)

Lamellar Membranes

Asif Hafeez and Zakariya Zubair

Abstract This book chapter gives a comprehensive review of two-dimensional (2D) nanomaterials that provide tremendous potential for separation applications. By using 2D nanomaterials as a building block, various types of lamellar membranes are actively investigated. In lamellar membrane, the mass transport occurs through the inherent interplanar spacing of the nanomaterials and the nanochannels generated during the fabrication of lamellar membranes. However, the trade-off between permeability and selectivity in membranes is inevitable. Different strategies such as interlayer channel aperture, membrane nanopores, and appropriate functionalization can bridge this trade-off. Various routes for preparing lamellar membranes including 2D nanosheet synthesis strategies, assembling techniques, mechanisms involved in transportation, and regulation of nanochannels are discussed in detail.

Keywords Lamellar membrane · Porous nanosheets · Non-porous nanosheets · Transport mechanism

1 Introduction

Separation of the target component from a mixture plays a vital function in various fields such as water/air purification, medical, chemical engineering, etc. Membrane technology is advantageous over conventional separation techniques (sorption, distillation, filtration, etc.) due to its benign environmental impact, low-cost, ease of fabrication, and operational performance [1–3]. Separation through membrane-based processes has received increased attention with polymeric membranes being in the foreground in research as well as the separation industry.

Membranes act as a barrier that selectively separates entities based on their nature and properties. The membrane acts as a selective barrier, permitting a species smaller than the pores through it while selectively separating one or more species larger than

A. Hafeez (✉) · Z. Zubair
Department of Materials, School of Engineering and Technology, National Textile University, Sheikupura Road, Faisalabad 37610, Pakistan
e-mail: ahafeez@ntu.edu.pk

K. Shaker and A. Hafeez (eds.), *Advanced Functional Polymers*, Engineering Materials,
https://doi.org/10.1007/978-981-99-0787-8_3

the pores. These species could be molecules of inorganic or organic entities. The efficiency of a membrane is translated in terms of permeability and solute exclusion and is based on the physio-chemical properties as well as pore shape, size and tunability through precise control. The membrane efficiency features such as high selectivity and high permeability while exhibiting stability in a harsh environment are still a challenge, which provides an avenue for the advancement of membrane technology.

The cost-effective synthesis and fabrication of a membrane that demonstrates exceptional stable separation features under harsh process conditions is an ongoing challenge. The trans-membrane channels are either composed of voids from dense polymer films, nanovoids from semi-rigid polymeric structures, or permanent voids from crystalline frameworks [4]. Ideally, the trans-membrane channels should be small with a high degree of compactness to obtain high permeance along with chemical and mechanical stability to ensure long-term application. The pore size distribution throughout the membrane should be narrow to provide membranes with higher selectivity [5].

Till now, several materials have been synthesized and used for gas and water purification to improve their limitations. The progress in the advancement of membranes based on polymeric materials is considered a milestone in the history of membrane technology due to its cost-effectiveness, efficient performance efficiency, and easy and facile fabrication at small to large scale. However, the polymeric membranes are prone to swell in organic solvents and exhibit lower stability at high temperatures, and are bound by the trade-off effect, where highly permeable membranes are generally demonstrating selectivity on the lower side and on the other hand, a membrane with low permeability exhibits a high selectivity [6, 7].

Recently, two-dimensional (2D) nanosheets with single- or a few-atoms thickness have become the anticipated structural block for futuristic nano-channeled membranes. Based on the atomic structure, 2D nanomaterials are either non-porous or porous. They can thus be assembled as a selective barrier for separation applications in the form of nanosheets and lamellar membranes, as shown in Fig. 1. The nanosheet membranes consist of a monolayer or a few layered 2D nanomaterial with uniform inherent pores or engineered nanopores for selective permeation. The lamellar membranes are fabricated by assembling 2D nanosheets overlayed with interlayer spacing used to provide molecular transport [8]. After the successful exfoliation of graphene by Novoselov and Geim in 2004, 2D nanosheets have gained widespread attention. The exfoliation resulted in a hexagonal lattice consisting of carbon atoms on an atomic scale [9]. Due to their thickness and unique structural characteristics, 2D nanomaterials present an opportunity for the development of membranes that can theoretically withstand ultimate membrane performance properties [10]. The reduced thickness of the membrane results in superior performance. The ultimate desirable thickness of the membrane would be one-atom-thickness which is achievable by two-dimensional (2D) materials [11]. Furthermore, 2D nanosheets can be designed with specific channel dimensions coupled with aimed chemical functionalities, which enables an exceptional membrane that is capable of physical or chemical selectivities [12]. Up till now, in addition to graphene, other two-dimensional materials such as zeolites, metal–organic frameworks (MOFs), covalent organic

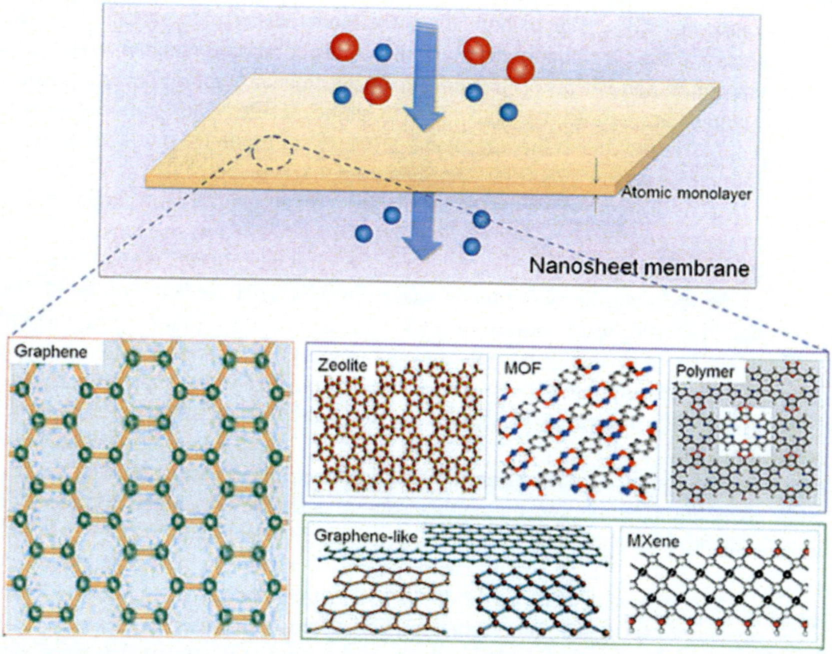

Fig. 1 Main Types of Two-Dimensional Materials and their Membranes. Reproduced with permission from [8]

frameworks (COFs), graphene oxides (GOs), hexagonal boron nitride, layered double hydroxides (LDHs), transition metal dichalcogenides (TMDs), MXenes and graphitic carbon nitrides (g-C_3N_4) have been employed for the fabrication of membranes [11, 13].

2 Lamellar Membranes

When 2D nanosheets are fabricated into membranes, ideally the single atomic thickness nanosheets with a high degree of porosity are mounted on top of the highly porous support [14]. The transmembrane short channels act as a transporting path for ultra-fast permeation. However, synthesizing such nanosheets with large intraplanar dimensions which are without any defects presents a challenge [15]. The lamellar membrane is another geometry that comprises prime quality 2D nanosheets as the main constituent and an impeccable membrane fabrication approach to fabricate a well-defined transmembrane pathway with abundant nanopores. These transmembrane channels provide a means of rapid transport to the penetrant. The mass transport takes place in both the intraplanar either inherent or engineered channels, and the interplanar channels. The transmembrane channels that are based on nonporous nanosheets originate from the interplanar spacing. The interplanar spacing channels are usually tortuous. These channel results when nanosheets are stacked parallelly, and due to the interplanar attractive or repulsive interactions. These interplanar interactions create an interlayer void that is responsible for mass transport [16, 17]. Two-dimensional nanosheets that are inherently nonporous includes GO, h-BN, MXenes, etc. Intrinsically porous nanosheets stacked in the lamellar fashion have additional nanochannels, that are established on the inherent pores of the nanosheets. Intrinsically porous two-dimensional nanosheets are based on zeolites, MOFs, COFs, and graphitic carbon nitride (g-C_3N_4). Nanochannels based on post-perforation can either be intrinsically porous or nonporous nanosheets. These intraplanar perforations within the nanosheets are introduced physically by charged particle beam or chemically through oxidation, or plasma to obtain engineered 2D nanosheets with enhanced porosity [10].

The lamellar assembling of 2D nanosheets has been acknowledged as a next-generation membrane for applications as an effective selective barrier due to its capability of being environmentally friendly as well as conserving space and energy [18]. Ideally, a high-performance membrane should exhibit outstanding durability and nanochannels that are more compact than the target molecules to maximize its permeance and solute exclusion. To realize a well-ordered 2D nanosheet lamellar structure, homogeneity of nanosheets with controllable physical aspects (shape, size, and thickness), the interfacial affinity between the polymer and 2D nanosheets is simultaneously essential. The resulting lamellar architecture would be responsible for molecular transport through interplanar slit-like nanopores and nanochannels, as shown in Fig. 2.

The current advancement of the 2D-enabled membranes with either inherent or engineered nanopores shows a unique pattern that varies from the traditional membrane designs. Up until now, three major classes of 2D nanosheet-based membrane systems have emerged with the advancement of two-dimensional nanomaterials, as shown in Fig. 3. The class 1 membranes (see Fig. 3a) are lamellar that are composed of closely stacked intermeshed nanosheet layers, where the nanosheets do not possess in-plane intrinsic pores. The convoluted and tortuous transport pathways are formed between the nanoflakes, also termed nanochannels and nanopores, that allow size-specific permeation of molecules and passage to the interplanar spacing between adjacent nanosheets of the packed two-dimensional nanosheets. Furthermore, the practical membranes in this class will exhibit a distribution of pore sizes and defects, originating from exfoliation. This range of pore size distribution is directly linked with nanoflake size distribution and degree of compaction. These nanoscopic transporting channels are termed external nanoporosity, which is illustrated in the highlighted area of Fig. 3a. Traditionally, Class 1 membrane represents nanoparticles with no intrinsic porosity such as graphene oxide [20] and h-BN [21] nanosheets. The membranes fabricated from Class 1 have been reported to efficaciously remove solvated solutes from the feed at exceptional permeability, suggesting the external

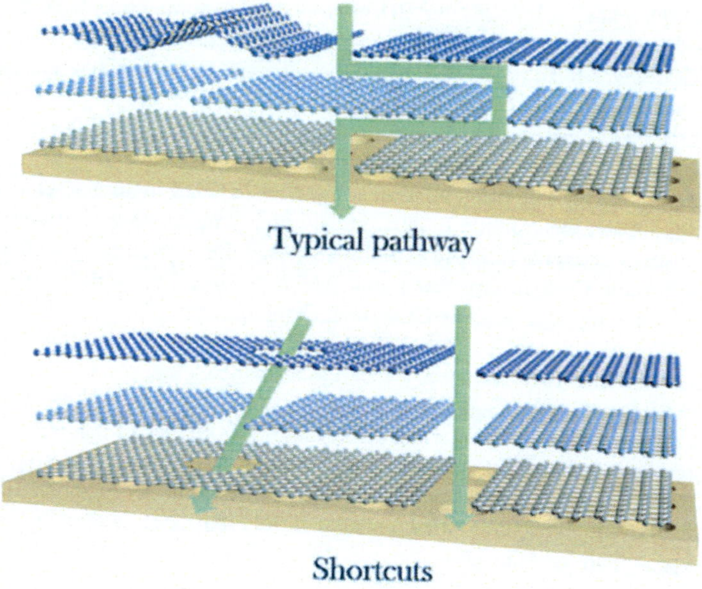

Fig. 2 Possible pathways for mass transport through a 2D lamellar membrane (top) interplanar mass transport (bottom) intraplanar mass transport. Reproduced with permission from [19]

nanoporosity ranges from < 0.7 nm to 0.7–2.0 nm [22] depending upon nanosheets and nanosheet stacking technique.

The nanomaterials with intrinsic basal plane nanopores used for two-dimensional membranes are classified as Class 2 membranes. The in-plane nanoporosity of these membranes is extremely capable of excluding molecular sizes up to 0.01–0.05 nm [23, 24]. These intrinsic nanopores are embedded in the three-dimensional (3D) porous nano-architecture and are different from external nanopores as well as intrinsic crystallographic defects. Nanosheets procured or delaminated for Class

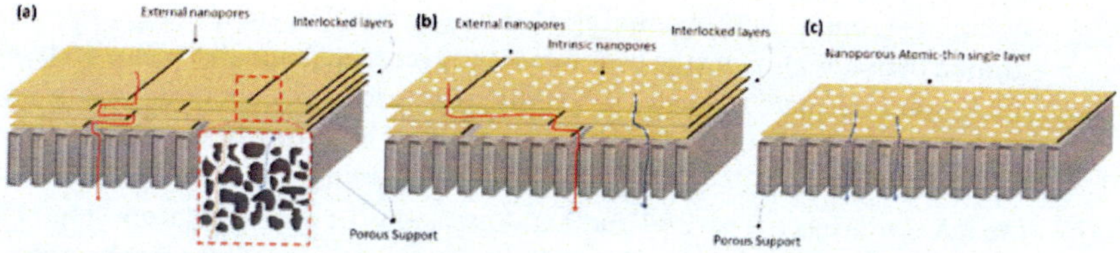

Fig. 3 Classes of 2D-nanosheet based membranes: **a** Class-1: membrane with intermeshed nanosheets, **b** Class-2: membrane with intermeshed nanosheets that have inherent nanopores, and, **c** Class-3: membrane with atomically thin layered nanosheet with inherent nanopores; red arrow indicates solvent transport through extrinsic nanopores and, blue arrow represent solvent transport through inherent nanopores

2 membranes originate from 3D microporous nanomaterials such as molybdenum disulfide (MoS_2) [25], zeolite [26], and, covalent- and metal–organic framework (COF and MOF) [24, 27]. Upon compaction of these intrinsic microporous nanomaterials into a lamellae formation, class 2 membranes exhibit synergistic enhanced size-specific permeation stemming from permeation promoting external nanopores and highly selective nanostructure-intrinsic nanopores, as shown in Fig. 3b. Recently, in-plane pore perforation in 2D nanomaterials is realized through elaborate oxidative etching and electron/ion bombardment employing microwave irradiation [28] and, plasma treatments [29]. Xu et al. (2018), used the oxygen plasma etching technique to introduce 0.6 nm in-plane pores, which enhanced the water permeation through reduced graphene oxide (r-GO) lamellar membrane.

The third class of 2D enabled membrane, see Fig. 3c, is an atomically thin 2D nanomaterial layer with induced or inherent in-plane nanopores. Class 3 membranes are an extension to the Class 2 membranes, yet they differ in thickness. Commonly class 2 membranes have a wider thickness ranging from tens of nanometres to micrometres due to nanosheet stacking. Nevertheless, class 3 membranes are distinguished as few-nanometre thick membranes assembled from 2D nanomaterials. Single- or few- atomic thick-layer membranes with inherent or engineered in-plane pores are representative of this class.

2.1 *Synthesis of Two-Dimensional Nanosheets*

A high-performance 2D nanosheet-based lamellar membrane requires superior quality 2D nanosheets with a precise membrane fabrication technique for the assembly and incorporation of nanochannels, that would provide a transportation pathway. Therefore, attention is required to the physical and chemical aspects of the nanosheets. The physical requirements of a nanosheet are its size, shape, and longitudinal and transverse dimensions. The chemical aspects that assist the transportation of entities either from nanopores or nanochannels are their defects and surface functionalities. These parameters dictate the selection criteria for fabrication techniques. The synthesis approach for 2D nanosheets can be generally divided into top-down and bottom-up routes, as shown in Fig. 4. The top-down technique signifies the disintegration of materials from the bulk into few-layered delaminated nanosheets, although the bottom-up route indicates the straightforward synthesis of 2D nanosheets from the corresponding starting materials. For a high-performance lamellar membrane fabricated, the nanosheets should be atomically thin to minimize the tortuosity of the mass transport. The longitudinal dimensions of the nanosheets should be wide to decrease the number of less selective intraplanar gaps. The nanosheets have intercalated moieties which facilitate scrupulous transport of penetrants in separation applications.

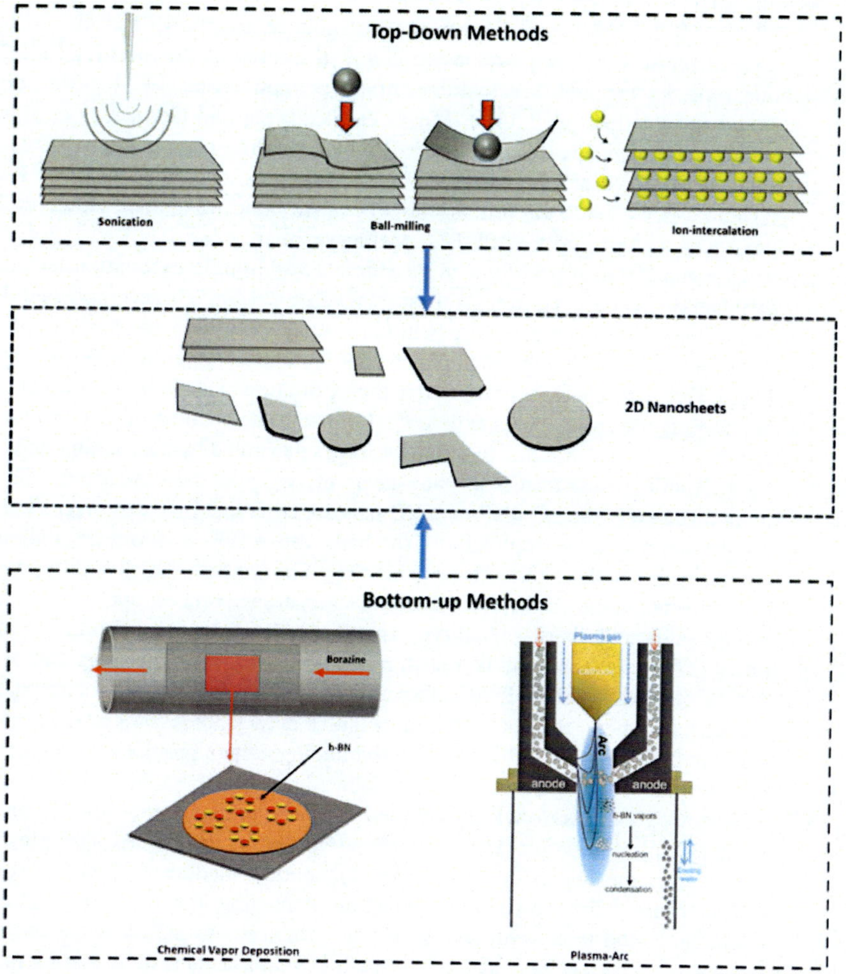

Fig. 4 Schematic illustration of 2D nanosheet synthesis techniques

2.1.1 Top-Down Route

2D nanosheets are assembled by stacking monolayers epitaxially. The layered structure of nanosheets is linked with strong covalent bonds within a lattice, while the weak interlayer interactions (vdW forces, H-bonding, and, π-π interactions) exist amongst the layers. As a result of the weak interlayer interactions, few-layered 2D nanosheets can be obtained from the bulk through the application of external force. The top-down methods are divided into solid-phase and liquid-phase exfoliation.

The disintegration of 2D nanosheets can be realized by solid-phase exfoliation using mechanical force. Single to few-layered nanosheets through microcleavage of the bulk layered crystals by peeling exfoliation technique. Exfoliation of bulk layered crystals is through adhesive tapes, in a study by Novoselov et al. (2004), they reported successful seclusion of graphene monolayers [9]. However, different studies by Pacile et al. (2008), Meyer et. al (2009), and Gorbachev et al. (2011) adopted mechanical exfoliation of h-BN by providing enough peeling force from adhesive tape to break the weak vdW forces, the process yielded very thin sheets of layered h-BNs, while the primary B-N covalent bond was intact. However, due to the stacking characteristic of h-BN, peeling exfoliation is not efficient [30].

In comparison to peeling a more efficient solid-phase exfoliation technique to disintegrate 2D bulk nanosheets through mechanical force is accomplished through ball-milling. The main principle behind the nanosheet disintegration from the bulk is the shear forces being delivered by the high-speed rotating balls, as shown in Fig. 4. The principal forces involved in the exfoliation of 2D bulk nanosheets through ball milling are shear forces and vertical impact. The disintegration of bulk layered crystals occurs majorly due to the shear forces generated by the rotating balls within the ball mill, which results in nanosheets of larger lateral dimensions. The other force involved in the exfoliation of 2D nanosheets is the force that originated due to the vertical impact and collision of the balls, this results in severe crystallographic defects with smaller lateral dimensions [31]. The vantage of ball-milling operation over other techniques is its scalability, convenience, and process optimization through milling parameters such as ball type and size, milling rpm, and ball/bulk nanosheet ratios [32]. This technique is a powerful process that introduces crystalline lattice defects and is even capable of destructing the crystalline lattice structure of the 2D nanosheets. Due to this introduction of lattice defects, the atoms around the defect sites are highly active and vulnerable to bonding, which provides an opportunity to introduce new functionalities to the exfoliated nanosheets [33].

Nowadays, the more widely adopted approach to disintegrate the 2D nanosheet from the bulk is by employing liquid-phase exfoliating techniques. These techniques are based on the dispersion of nanosheets in a liquid medium followed by the application of mechanical forces or introduction of an entity that disrupts the interplanar interactions and then applying the mechanical force to exfoliate the few-layered nanosheets. The mechanical force to the bulk nanosheets is delivered by ultrasonication technique. The colloidally stable exfoliated 2D nanosheet that is comprised of small and few-numbered stacked layers requires further separation by sequential centrifugation to yield the exfoliated nanosheets. The liquid-phase ultrasonication-based exfoliation of nanosheets is favored due to its process optimization through sonication intensity, amplitude, and period. A strategic solvent selection is also crucial for effective exfoliation, Exfoliation by sonication is effortless when the difference in surface energy of the solvent and nanosheets is marginal, hence the smaller mixing enthalpy. Therefore, the surface energy of the solvent plays a vital role in optimizing the sonication conditions. However, the surface energy is primarily dependent upon the temperature, which alters with the sonication intensity and duration. The surface energy of the solvent can be further enhanced by using a co-solvent or using a solute.

The liquid-phase exfoliation of 2D nanosheets which is assisted through intercalating ions or molecules into the layers is another route of liquid-phase top-down exfoliating nanosheets from the bulk, as illustrated in Fig. 4. The basic principle behind this technique is to attach ions of the small ionic radius into the interplanar spacing of the layered nanosheets. The ion attachment results is a significant increase in the interplanar spacing, thus weakening the vdW interactions amongst the consecutive layers in the layered bulk crystals. The intercalation of external moieties functions either by weakening the layer-to-layer interactions or by enhancing the surface polarity which improves the solvent dispersion. The ion-intercalated 2D nanosheets can be readily exfoliated into single/few-layered nanosheets under gentle sonication conditions. As a typical example various cationic entities such as Li^+, K^+, Zn^{2+}, Na^+, etc., are intercalated in the interlayers of the bulk crystals through forced hydration. In most cases, the ion-intercalated 2D nanosheets interact with the solvent and generate hydrogen gas upon reaction, which can also assist in expanding the layers in close vicinity during the sonication process and therefore results in promoting the exfoliation efficiency [34]. Molybdenum disulfide (MoS_2) and, transition metal dichalcogenides (TMDs) are usually exfoliated by this technique. This intercalation exfoliation technique enables a highly stable dispersion of 2D nanosheets in a different solvent. However, the prerequisite of this technique requires a longer time and smaller nanosheet dimensions for proper diffusion of ions within the interlayers [34]. Further purification by removal of large and thick flakes by centrifugation enhances the yield of single- or few-layered nanosheets.

2.1.2 Bottom-Up Route

The top-down synthesis route is limited to the presence of layered nanomaterial as the predecessor, whereas the bottom-up technique is related to almost all 2D nanomaterials. This technique involves the build-up of materials from scratch (atom-by-atom or cluster-by-cluster), generating a uniform nanosheet size, contour, and, distribution. Bottom-up synthesis techniques are mainly comprised of wet-chemical synthesis and chemical vapor deposition (CVD) growth. These processes are equipped to precisely control the synthesis process to maintain the nanoparticle size by inhibiting the reaction process. Furthermore, these techniques surpass in regulating the quantity and quality of the nanosheets with homogenous size distribution and morphology.

The wet-chemical synthesis is employed for the synthesis of the typical porous 2D-nanosheets that are made from nodes and linkers like zeolites, MOFs, and COFs. The synthesis process selectively inhibits the growth of the material unidirectionally [35]. Hydrothermal, interfacial synthesis, and solvothermal techniques are the most popular strategies. These techniques involving the synthesis of 2D-nanosheets in aqueous and solvents are often carried out under high temperatures and pressures [34]. The lateral and transverse dimensions as well as the size and degree of crystallinity of the 2D nanosheets can be precisely controlled through meticulously changing the synthesis parameters and by utilizing surfactants and structure-directing agents. The interfacial synthesis applies to thin-film membranes (TFN) as well as nanosheets. In this method, the low thickness is insured through the reacting monomers in a confined interface. The interfaces include liquid–liquid, liquid–gas, or liquid–solid. The structural integrity of the produced 2D nanosheets and TFN membranes is regulated by the diffusivity of the monomers towards the interface [36].

CVD is a proficient and effective technique to synthesize high-quality 2D nanoparticles and few-layered films of atomic thickness on a solid substrate. In CVD, the precursors gas or vapor are reacted or disintegrated on the substrate at high temperature in a vacuum chamber, as depicted in Fig. 4. In this technique, 2D nanosheets augment the substrate in the absence or presence of a catalyst. CVD have been employed for the epitaxial growth of various 2D-nanosheets, such as transition-metal dichalcogenide (TMD) [37], metal carbides [38], graphene [39], borophenes [40], silicene [41] and, h-BN nanosheets [41, 42]. The CVD method allows the production of highly crystalline 2D-nanosheets with desirable longitudinal and lateral dimensions, for being potentially utilized as membrane precursors.

2.2 *Preparation of Nano-Channeled Lamellar Membrane*

2D nanosheets with atomic thickness can be employed as a means of separation in membranes. The presence of pores is provided either by 2D nanosheets that have an inherent porous structure such as in zeolites and MOFs or by drilling pores such as in nanosheets like graphene. The performance of membrane-based nanosheets in terms of permeance and selectivity is directly related to the thickness and porosity of the 2D nanosheets. These performance characteristics are predominantly defined by the selection of high-quality synthesis routes and an effective approach for transforming these 2D nanosheets into a membrane. This section focuses on the preparation of 2D nano-channeled membranes.

2.2.1 Development of Single-Layered Membranes

The single-layered film with an atomic thickness anchoring nanopores is regarded as an ultimate membrane. However, realizing such membranes presents an immense challenge due to their crack-free transformation into large-area nanosheets with the ability to control their pore generation. Graphene is deemed an ideal candidate for a single-layered membrane while factoring in the requirements of a 2D nanomaterial which have the potential of scalability while exhibiting robust mechanical properties. However, membrane fabrication defects like cracks, tears, or wrinkles are usually evolved during transferring graphene onto a porous substrate, which results in poor membrane separation performance [17]. To rectify the membrane fabrication defects two different approaches are adopted. Firstly, a common solution is to coat graphene with a reinforcing polymer i.e., polymethylmethacrylate (PMMA) to improve its mechanical durability [43]. However, the polymeric coating incapable of permeating must be removed in the post-processing of the single-layered membrane, which inevitably introduces cracks. The second approach to remedy these fabrication defects is to use a porous substrate material instead of a nonporous polymeric material. The porous substrate material should have a preferential affinity towards graphene which assists in a more effective large-area transfer [44]. Although considerable advancements have been made in single-layer 2D nanosheet membranes, the scalability of these membranes is difficult due to the introduction of membrane fabrication defects. These difficulties can be partially relieved by increasing the number of layer stacking by compromising the membrane permeance. Furthermore, eliminating the transfer process is another aspect of single-layer 2D nanosheet membrane fabrication. This is realized by bottom-up direct and *in-situ* development of continuous single-layer porous nanosheet membranes on a porous substrate.

2.2.2 Development of Lamellar Membranes

Two-dimensional nanosheets which manifest stable nanosheet suspension in a solvent being deemed as solution-processable can be easily transformed into a lamellar structure. The prerequisite for two-dimensional nanosheet transformation to the lamellar membrane is its ability to be solution processable. Furthermore, 2D nanosheets should exhibit excellent solvent dispersibility to materialize the nanosheet suspension in different solvents. This means nanosheet dispersion is readily processable into a lamellar structure through traditional casting or coating techniques [17]. Lamellar membrane fabrication with 2D nanosheets in-plane to the support has precedence in terms of reducing hydrodynamic resistance through deposition thickness and generation of adequate through-plane channels. Ideally, the lamellar membrane should hold being aligned parallelly and with intermeshed layers. Various techniques like filtration, coating, and layer-by-layer (LBL) deposition have been used to fabricate 2D nanosheet lamellar membranes, as shown in Fig. 5.

Filtration

The positive and negative force applied to filter through substrate via pressure and vacuum is one of the most frequent and effective deposition techniques, as shown in Fig. 5a. This vertical downward push, provided by positive/negative force drives the nanosheet suspension into an interlocked layered structure on the substrate [27]. Class 1 (Fig. 3a) and Class 2 (Fig. 3b) nanosheet membranes are most fabricated using this technique. The deposition can be easily modulated from nanometre to micrometre scale by varying the deposition concentration. This technique offers the liberty to intercalate other dispersible nanoparticles or ions during the deposition to regulate the nanochannel structure [45]. Chen et al. (2019), developed h-BN based lamellar membrane through vacuum filtering of amine-functionalized nanosheets. The thickness through deposition is adjustable through varying concentrations of

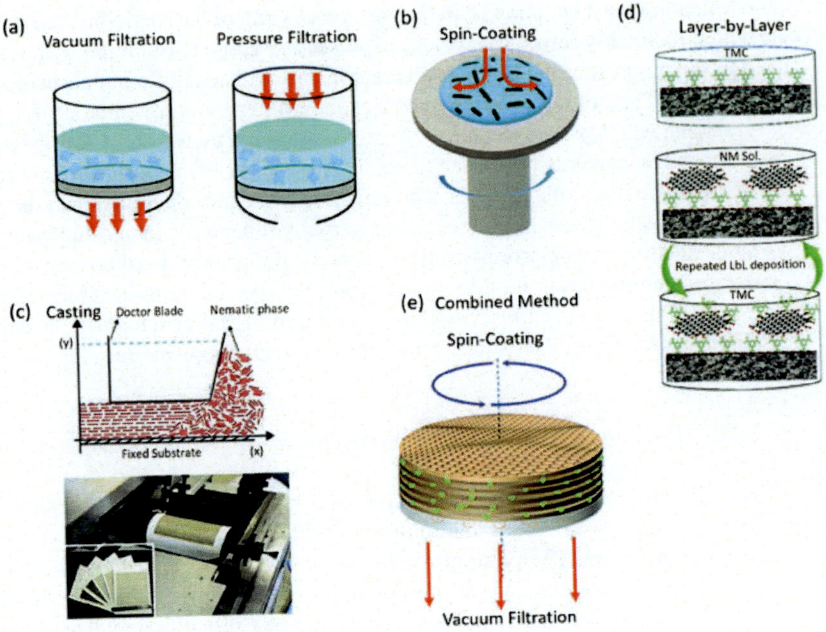

Fig. 5 Schematic illustration of lamellar membrane fabrication methods: **a** Vacuum/Pressure filtration, **b** Spin-coating, **c** Casting, **d** layer-by-layer deposition and **e** combination of spin-coating and vacuum filtration [17]

amine-functionalized h-BN suspension. The fabricated lamellar membranes demonstrated excellent water and organic solvent permeability while excluding various cationic as well as anionic dyes [46]. Recently, Lin et al. (2020), crosslinked activated h-BN and GO to assemble an a-BN/GO hybrid membrane that demonstrates remarkable exclusion efficiency as well as resilience in liquid separation. The compact lamellar structured hybrid membrane was fabricated through pressurized filtration. The a-BNGO membrane exhibited 99.98% methylene blue rejection with water permeance of 4.15 L m^{-2} h^{-1} bar^{-1} [47].

Coating

There are different coating techniques used to assemble 2D nanosheets into lamellar membranes. The most common technique employed for this fabrication method involves spray-coating, casting, spin-coating, and dip-coating. Uniform assembly and nanosheet distribution of the coating depends upon the nanosheet-solvent interaction (surface tension of the solution) and, interfacial properties of the nanosheet suspension (rate of evaporation) and substrate (smooth surface). The spray-coating method involves spraying a dilute yet volatile nanosheet-solvent suspension onto the support. The evaporation of the solvent leaves a self-assembled pseudo-lamellar structure on the support. The thickness of the coating can be maneuvered by changing the suspension viscosity and spray pressure. Casting is also categorized as under-coating and has been vastly applied for polymeric membrane fabrication. The casting method is easily controllable and scalable. This technique has also been reported to fabricate 2D-nanosheet laminar membranes, as shown in Fig. 5c. The shearing forces generated by the transverse movement of the doctor-blade assists the nanosheet alignment in the cross-direction. In the spin-coating method (Fig. 5b), the nanosheet-solvent suspension is poured onto the center of the support and is evenly distributed through centrifugal force. This technique enables a thin, evenly distributed, and highly interlocked nanosheet lamellar membrane. Lastly, dip-coating involves dipping the support membrane into the nanosheet suspension and drawing it out while allowing the solvent to evaporate and form a self-assembled lamellar structure. This technique is majorly employed for Class 1 2D-enabled membranes. An et al. (2019), transformed the hydrophobic surface of polyvinylidene fluoride (PVDF) into a superhydrophobic surface by incorporating functionalized boron nitride nanosheets in the dope solution. The composite functionalized h-BN and PVDF membrane were fabricated through the solution casting method and phase-inversion technique. The composite membrane shows excellent performance for oil separation from an oil–water mixture [48].

Layer-By-Layer (LbL) Assembly

This is the method of alternatively depositing materials of different characteristics on the substrate surface, Fig. 5d. This technique primarily depends upon the interaction between the alternating and adjacent layers. The interaction between the adjacent layers could be electrostatic, vdW forces, H-bonding, or even covalent bonding. Layer-by-Layer (LbL) lamellar membrane fabrication technique presents precedence over other fabrication methods in precisely controlling the deposited thickness of the selective layer by changing the cycles of deposition. However, this technique requires 2D nanosheet compatibility in various polar solvents as well as the adjacent layer material. Abdikheibari et al. (2019), established a modified LbL technique for the assembly of highly permeable, highly selective yet fouling resistant thin-film membrane being decorated with few-layered amine functionalized h-BN nanosheets for the separation of hydrophilic natural organic matters. In their fabrication process, h-BN was covalently bonded to the adjacent layer of the membrane [49].

Combined Method

At present, lamellar membrane preparation solely depends on a single method. The combined tuning of nanosheet alignment in the vertical and horizontal directions is difficult. The filtration through pressure and vacuum can align the nanosheets vertically, while the casting and coating technology has the capability of aligning nanosheets horizontally. A merger of processes, as shown in Fig. 5e, where filtration is combined with spin casting could allow tuning of stacking both vertically as well as horizontally. This technique employs a combination of filtration and spin-coating technique that was reported to fabricate GO lamellar membranes which resulted in a well-detailed 2D-nanosheet lamellar membrane with tailored intraplanar and interplanar nanochannels [50].

The scalability of the membrane preparation technique is essential in the implementation of practical applications. The techniques involving casting and coating that are easily scalable and can be adopted at an industrial level production of thin lamellar flat sheet membranes. Moreover, the deposition and attaching of the 2D nanosheets to the flat sheet and hollow fibers through the pressure-filtration technique could assist in scale-up. It is worth mentioning that all the processes are fundamentally dependent upon the uniform suspension of these 2D nanosheets in a solvent. However, the necessary modifications to enhance the dispersion of 2D nanosheets in a lamellar membrane fabrication technique appropriate solvent is the most important prerequisite.

3 Mechanism of Transportation

In the membrane exclusion process, penetrants pass through the membrane based on their size and the membranes serve as a selective barrier. Membranes achieve this selective transport through various mechanisms, which are dependent upon the pore size, see Fig. 6. Furthermore, the smallest pore size scale ranges from 0.3–0.6 nm, corresponding to dense membranes having no specific pores such as reverse-osmosis membranes for water desalination and various gas separation membranes utilizing the solution-diffusion mechanism [51]. The selectivity and exclusion of the solute result from solubility as well as diffusivity differences in the membrane material. The solubility of the solute is dependent upon its molecular structure, the porosity of the membrane, and the chemical affinity, whereas the solute's ability to diffuse in the membrane material is controlled by the rearrangement of the polymeric chains which are activated thermally that favors size-based diffusion of smaller molecules. When the pores are wider than the molecular size but smaller than the gas mean free path, the transmembrane transport is governed by Knudsen diffusion [5].

The pathway for mass transportation across a 2D lamellar membrane depends upon the in-plane porosity, as shown in Fig. 7c. For a lamellar membrane fabricated from intrinsically porous nanosheets, the pathway for transportation includes the pathway through nanochannels i.e., the interplanar spacing of neighboring nanosheets and nanopores i.e., intraplanar porous structure. In addition, the lamellar membranes fabricated from the 2D nanosheets that do not have inherent or perforated nanopores present a pathway through the interplanar spacing between the consecutive nanosheets. The permeance through the membranes is dependent upon lamellar membrane porosity and tortuosity. In particular, the porosity is associated with nanosheet interplanar spacing, the thickness of the nanosheets, and the pore

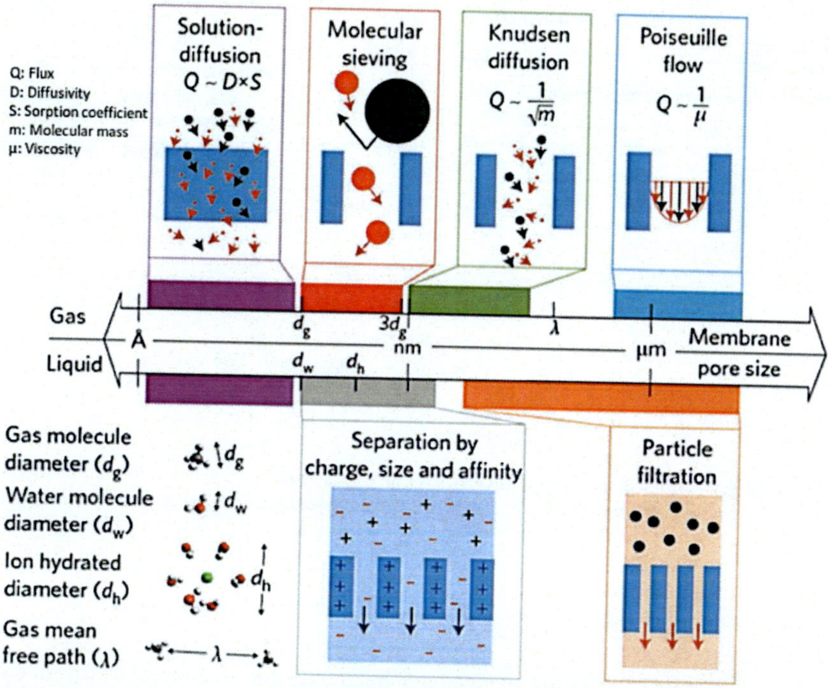

Fig. 6 Membrane transport mechanisms based on the membrane pore size. Reproduced with permission from [5]

size of the intrinsic or perforated nanosheets, while the tortuosity is a ratio of lateral dimensions to the thickness of the 2D nanosheets. In the interplanar pathway, as shown in Fig. 7c, the length of diffusion can be estimated by the Nielsen transport model [52].

$$l = h + N{}^{L}\!/_{2},$$

$$N = \frac{h}{d + W},$$

where l is diffusion length, h is the lamellar membrane thickness, L and W are the lateral length and thickness of the nanosheet, and d is the basal spacing of the nanosheets. In this section, the transportation mechanism of lamellar membranes fabricated through 2D porous and nonporous nanosheets is discussed.

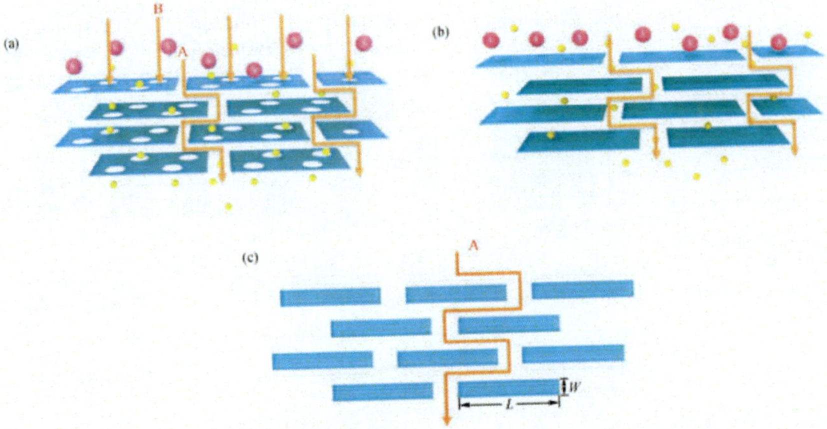

Fig. 7 Molecular transport model for **a** Intrinsically/post-perforated porous 2D nanosheet lamellar membrane, **b** Nonporous 2D nanosheet lamellar membrane, and, **c** Interplanar pathway. Reproduced with permission from [11]

3.1 *Porous Nanosheet-Based Lamellar Membranes*

The specific molecular exclusion of inherently porous nanosheets-based membranes is dependent on the size of intraplanar pores, as shown in Fig. 7. Due to the size of the nanopores, intrinsically porous nanosheets-based lamellar membranes will permeate and exclude ions or molecules based on size, see Fig. 7a. According to the nanopores and nanochannel size, the molecular diffusion across the membrane is derived by surface, molecular, and Knudsen diffusion and viscous flow [53]. The lamellar membrane fabricated through stacking intrinsically porous nanosheets exhibits a regular nano-porous structure coupled with an ultrathin thickness which presents an ideal separation membrane morphology, exhibiting limited hydrodynamic resistance to the mass transfer and promoting very high flux and maintaining high solute exclusion resulting from precise pore sieving.

3.2 *Non-Porous Nanosheet-Based Lamellar Membranes*

Two-dimensional nanosheets without intrinsically in-plane nanopores stacked to form a lamellar membrane resulting in the formation of interplanar nanochannels that provide a mean for the transportation of ions or molecules, as shown in Fig. 7b. The interplanar spacing between the 2D nanosheets between the consecutive nanosheets provide a pathway of transportation for small entities whilst restricting the flow of entities that are larger than the nanochannels. For 2D lamellar membranes that are based on non-intrinsic pores, the development and modulation of the interplanar spacing is a fundamental design characteristic for membrane performance in terms of permeance and solute exclusion.

3.3 Transportation Through Interlayer Channels

Two-dimensional nanosheets when stacked by the lamellar membrane fabrication techniques result in the creation of interlayer nanochannels. The interlayer spacing in 2D- nanosheets based membranes is dictated by the pathways amongst the nanosheets. These channel size and the chemical functionalities present on the nanosheets dictates the separation efficiency of these membranes [12]. As shown in Fig. 8a, the larger ions and molecules that are hydrated are stopped through a lamellar membrane assembled using GO nanosheets while permitting smaller ions and molecules to transport through these channels. The separation mechanism of these hydrated ions and molecules can be explained with the help of size exclusion theory [54]. Thin layered nanosheet films fabricated by chemically exfoliating WS_2 nanosheets have been reported to separate 3 nm small particles. Furthermore, the permeance of the lamellar membrane could be enhanced through pressure-assisted cracks and utilizing templates, which generate more fluidic channel networks, as shown in Fig. 8b [55].

The development mechanism of nanochannels can be associated with interplanar interactions that include non-covalent and covalent interactions. The nanochannels can be modulated by changing these interactions. These non-covalent interactions

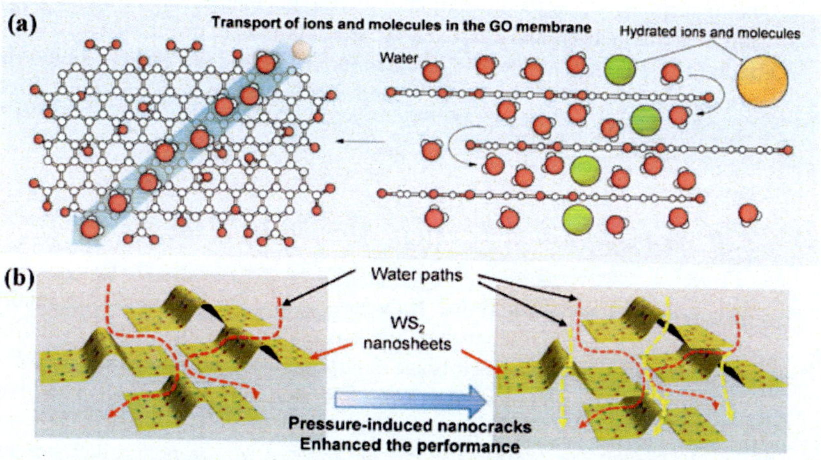

Fig. 8 **a** Water and small-sized ions and molecules are being permeated through stacked GO nanosheet superfast membrane, while inhibiting the larger molecules, **b** Illustration of ultrafast molecule separation through layered WS_2 nanosheet membranes and its enhanced permeance due to the introduction of pressure-induced nano cracks. Reprinted with permission from [56]

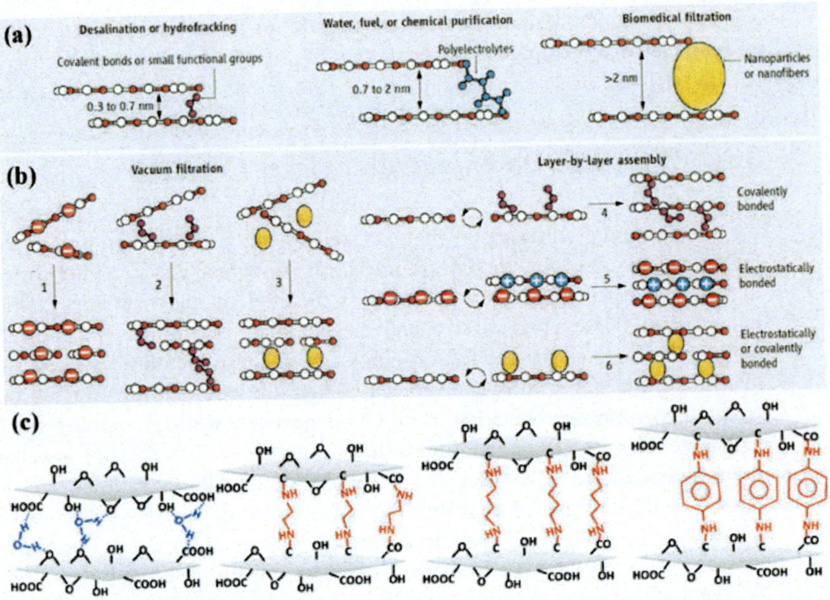

Fig. 9 **a** The membrane performance of the GO membrane is tunable through modulating interplanar spacing, **b** Different approaches for the fabrication of GO nanosheet-based membrane; (options 1–3) GO nanosheets stacked through vacuum filtration, (options 4–6) nanosheets being stabilized through covalent, electrostatic or combination of covalent and electrostatic bonding

being responsible for the formation of interlayer channels are further segregated into physical interactions and internal and external forces. The internal forces refer to the secondary bonding at play in adjacent monolayers, and these forces include hydrogen bonding, hydrophobic effects, vdW, and π-π interactions. The internal forces play a vital role in holding the multilayers together.

3.4 *Transportation Through Membrane Nanopores*

The driving force behind the transport of entities through a nanoporous 2D-nanomaterial-based lamellar membrane is mainly dictated by the pore dimensions. As discussed in Sect. 3.0, membrane nanopores allow entities with a smaller diameter than the pore size and block the larger hydrated ions and/or molecules. The dimension of nanoporous 2D nanosheet pores can be modulated during the synthesis. In a typical top-down synthesis, pores are created from a nonporous atomically thin material, and from bottom-up synthesis, the nanoporous structure can be created through intrinsically porous materials such as 2D-MOF, 2D-COF, and 2D-polymers [5].

In a top-down synthesis, a homogenous distribution of sub-nanometer pores can be achieved through the utilization of ion irradiation or chemical/plasma etching. The chemical processes influence the stability of the pores, and the functional groups present in the vicinity of the pores. The size and the density of the nanopores are dictated by the interplay between the nucleation of the defects and their growth. Thermal oxidation, oxygen, and hydrogen plasma are capable of producing uniform nanopores with tunable size and density [57]. Furthermore, nanopore generation through ion irradiation is realized where the defect nucleation is controllable. The nanopore optimization in terms of size and density is achieved by changing the type of ions, their angle, and energy. In addition, a high density of sub-nanometer pores is also achievable by electron-induced sputtering or chemical etching.

In the bottom-up synthesis techniques, intrinsically porous materials are the building blocks for a well-detailed porous structure and a high density of atomically precise pore size [58]. Cheng et. al., fabricated a two-dimensional MOF nanosheet lamellar membrane through the spin coating method. The membrane exhibited a precise pore size of 0.52 nm, which allows the permeation of smaller gas molecules (CO_2, kinetic diameter of 0.330 nm), while the membrane presents resistance to larger gas molecules (CH_4, kinetic diameter of 0.376 nm) [59].

4 Nanochannel Architecture and Its Regulation

A nanochannel is the building block of 2D nanomaterial-based membranes, which connects the membrane fabrication and its separation process. The separation performance of the membrane is determined by the architecture and microenvironment of the nanochannels. The performance of the membrane can be fine-tuned through the optimization of nanochannel architecture and its surrounding environment.

4.1 Surface Functionalization

The selectivity and permeability tradeoff of membranes has always been a challenging problem. The selectivity and permeability of 2D nanomaterial-based membranes are not completely dependent on the interlayer channels and nanopores in the 2D nanomaterials, which utilize the interplanar electrostatic interactions and size exclusion mechanisms. Meanwhile, advancements in the development of 2D nanosheet-based lamellar membranes have been achieved by adopting versatile approaches for the modification of interplanar channels or nanopores. Modulated pore size, adjustable inter and/or intraplanar electrostatic interactions, and improved solid–liquid and solid–gas interfacial properties have been made possible with the surface functionalization of the 2D nanosheet-based lamellar membranes. The gap between the selectivity and permeability tradeoff can be bridged to some extent through the 2D nanosheet surface functionalization.

Inspired by the applications of 2D nanosheets-based lamellar membranes in ions or molecules separation various theoretical studies simulated the transport performance of 2D nanomaterial materials through molecular dynamics simulations [60, 61]. Konatham et al., studied the transport of water and ions through the nanopores of porous graphene nanosheets by MD simulation [62]. The simulation results suggested that the nanopores are tunable with the introduction of various charged surface functionalities. The diameter of graphene nanopores with charge groups OH^-, COO^-, and NH^{3+} were 0.75, 1, and, 1.1 nm, respectively. It was found that the OH^- functionalized nanopores with a narrow diameter of 0.75 nm were highly stable in excluding Cl- ions even at a moderate ionic strength of the solution. Experimentally, Hafeez et al., functionalized h-BN nanosheets with different molecular weight polyethylene glycol and fabricated lamellar membrane through filtration on a substrate membrane [63]. The study revealed interplanar modulation through soft spacer molecules.

4.2 Nonporous 2D Nanomaterials

A lamellar membrane fabricated through stacking nonporous 2D nanomaterials is primarily separated by the interplanar spacing which provides a means of transport. The formation mechanism of nanochannels is closely associated with non-covalent and/or covalent interlayer interactions. These non-covalent and covalent interactions can be adjusted to regulate the nanochannels. The non-covalent interactions are further categorized into intrinsic interactions (hydrogen bonding, hydrophobic effects, van der Waals and π-π interactions), and the forces that are applied during the nanochannel formation or operation that are referred to as extrinsic interactions (centrifugal force, pressure, etc.).

In 2D nanomaterial-based lamellar membranes, the interplanar interactions such as van der Waals, π-π interactions, hydrogen bonding, etc., are referred to as internal forces which hold the adjacent layers. The intensity of these internal forces can be regulated through various pre or post-thermal or chemical treatments, which alter the aperture of the interplanar nanochannels. The homogenous distribution of functional groups on the nanosheets will result in uniform interplanar interactions which would bring a well-ordered structure. Uniform distribution of functional groups on the nanosheets enables a regular lamellar structure, MXenes are decorated with homogenous abundant surface-terminating groups as compared to GO which has randomly distributed oxygen groups. The regular free spacings generated by interplanar secondary forces could perform as an effective nanochannel that separates based on the size exclusion mechanism [64]. The oxygenated groups on GO prevent interplanar restacking through π-π interactions. The π-π interactions in GO can be enhanced through thermal annealing or chemical reduction, which would allow interplanar restacking. Furthermore, these post-treatments would produce robust and narrow apertured nanochannels [12].

During 2D nanosheet deposition, external forces such as positive or negative pressure, shear forces, etc., can be modulated to orient the nanochannels, which provides effective control over the fabrication process. Two-dimensional h-BN and GO nanosheets are typical examples of lamellar membranes prepared by pressure, vacuum filtration, or evaporation which results in ordered, partially ordered, and random ordered nanochannels respectively [13, 63, 65]. The evaporation and drop-coating methods predominantly rely on the slow evaporation of the solvents to form a lamellar structure, which results in a lamellar structure with nanochannels that are less ordered [66]. The nanochannel orientation can be further improved through the application of shear force which creates an interlocked nanochannel structure. Furthermore, the combination of pressure exertion in the vertical direction and shear forces in the horizontal direction produces a highly ordered nanochannel structure [50]. Besides the type of force, the magnitude of forces provides a higher tunability of the nanosheet stacking mode.

For 2D nanosheets that can be covalently bonded, chemical intercalation through crosslinking is the most common route to enhance the interplanar forces and improve the mechanical stability of the membranes. The intercalating agents encompass charged entities, small molecules, or macromolecules. GO is the most explored 2D nanosheet for chemical intercalation considering the abundant functional groups. Diamines are versatile intercalating agents of GO, which is performed through nucleophilic substitution reaction between epoxy and amine. The aperture of nanochannel can easily be regulated by changing the chain length of diamines [67, 68]. The lamellar membrane application varies from gas separation, water purification, and organic solvent filtration. Covalent crosslinking requires the presence of anchoring functional groups on the nanosheets, and strong covalent interactions can ensure structural integrity to avoid swelling in liquid filtration. These intercalated molecules not only serve as a means to fine-tune the interplanar spacing but can also introduce a chemical microenvironment in the nanochannel that is penetrant-philic, which facilitates targeted molecule transportation.

4.3 Intrinsically Porous 2D Nanosheets

The nanochannels for the mean of transportation in intrinsically porous 2D nanosheets are the nanosheet interplanar spacing and the intrinsic porosity of the nanosheets. This intrinsic porosity presents an opportunity for building a selective transport pathway. This versatility is provided by 2D nanosheets of zeolite, metal-covalent organic framework (MOF), and covalent-organic framework (COF). The majority of 2D zeolite nanosheets have a narrow pore size distribution of 0.2–1 nm, which is close to the molecular sizes of most of the gases. However, depending upon the organic linkers of MOFs, they have a wide pore size distribution ranging from 0.3 to 10 nm, which makes them a suitable candidate for gas and liquid separations. Furthermore, COFs have a pore size distribution ranging from 0.8 to 5 nm. Due to the high degree of crystallinity, intrinsically porous 2D-nanosheets are brittle, which makes its single-layer membrane processing challenging and as a result, structural defects are introduced [36]. The nanochannel modulation based on the penetrant molecular size is mainly focused on the pre-design of nods and linkers of the starting intrinsically porous 2D nanosheets [66]. The aperture of the nanochannel and the chemical functionality present can be easily manipulated.

4.4 *Perforated 2D Nanosheets*

Pore generation on a nanoscale in 2D nanosheets is another important nanochannel regulation strategy. The prerequisite for pore generation in 2D nanosheets through peroration is that the nanosheet exhibits inherent superior mechanical strength. The various approaches used for perforation include physical techniques with focused electron beam irradiation, focused ion beam irradiation, and ion bombardment and chemical techniques involving chemical oxidative etching, plasma, and ultraviolet (UV) light etching. The perforation of nanosheets using focused electron beam irradiation uses a high-energy electron beam to etch carbon atoms and induce structural rearrangement, forming nanopores in the range of 5 to 100 nm on the graphene. The degree of perforation from the micron to the nanoscale can be optimized by controlling voltage, exposure time, and beam dosage. The chemical perforation techniques involve the introduction of sp^3 hybridized sites or vacancy defects. In these techniques, carbonyl or hydroxyl functional groups are introduced through oxidative processes forming nanochannels with nanoscale precision [69]. The nanopore generation through perforation is capable of high precision over pore size and density.

5 Conclusion

The emergence of various 2D nanomaterial-based lamellar membranes offers a unique opportunity for developing ultra-thin nanochannel membranes which cater to many fields, ranging from daily life to medical science. In terms of transportation routes, 2D nanomaterials present two ways for the separation of solid–liquid or solid–gas interaction. Meanwhile, to unravel the selectivity-permeability tradeoff, interlayer functionalization has been considered one of the most viable options which improves the microenvironment for the penetrant molecules. The vast diversity of synthesis routes, nanosheet functionalizing, and nanochannel modulation enables the application of 2D nanomaterials in application-specific membranes.

References

1. Zhang, R., Liu, Y., He, M., et al.: Antifouling membranes for sustainable water purification: strategies and mechanisms. Chem Soc Rev **45**, 5888–5924 (2016)
2. Joshi, R.K., Alwarappan, S., Yoshimura, M., et al.: Graphene oxide: the new membrane material. Appl Mater Today **1**, 1–12 (2015)
3. Li, W., Zhang, Y., Li, Q., Zhang, G.: Metal- organic framework composite membranes: Synthesis and separation applications. Chem Eng Sci **135**, 232–257 (2015)
4. Galizia, M., Chi, W.S., Smith, Z.P., et al.: 50th anniversary perspective: polymers and mixed matrix membranes for gas and vapor separation: a review and prospective opportunities. Macromolecules **50**, 7809–7843 (2017)
5. Wang, L., Boutilier, M.S.H., Kidambi, P.R., et al.: Fundamental transport mechanisms, fabrication and potential applications of nanoporous atomically thin membranes. Nat Nanotechnol **12**, 509–522 (2017). https://doi.org/10.1038/nnano.2017.72
6. Robeson, L.M.: The upper bound revisited. J Memb Sci **320**, 390–400 (2008)
7. Caro, J.: Hierarchy in inorganic membranes. Chem Soc Rev **45**, 3468–3478 (2016)
8. Liu, G., Jin, W., Xu, N.: Two-Dimensional-Material Membranes: A New Family of High-Performance Separation Membranes. Angew Chemie - Int Ed **55**, 13384–13397 (2016). https://doi.org/10.1002/anie.201600438
9. Novoselov KS, Geim AK, Morozov S V, et al (2004) Electric field effect in atomically thin carbon films. Science (80-) 306:666–669
10. Koenig, S.P., Wang, L., Pellegrino, J., Bunch, J.S.: Selective molecular sieving through porous graphene. Nat Nanotechnol **7**, 728 (2012). https://doi.org/10.1038/nnano.2012.162
11. Wang, W., Wei, Y., Fan, J., et al.: Recent progress of two-dimensional nanosheet membranes and composite membranes for separation applications. Front Chem Sci Eng **15**, 793–819 (2021). https://doi.org/10.1007/s11705-020-2016-8
12. Kim HW, Yoon HW, Yoon S-M, et al (2013) Selective gas transport through few-layered graphene and graphene oxide membranes. Science (80-) 342:91–95
13. Hafeez, A., Abdul, Z., Fauzi, A., et al.: Functionalized boron nitride composite ultrafiltration membrane for dye removal from aqueous solution. J Memb Sci **612**, 118473 (2020). https://doi.org/10.1016/j.memsci.2020.118473
14. Celebi K, Buchheim J, Wyss RM, et al (2014) Ultimate permeation across atomically thin porous graphene. Science (80-) 344:289–292. https://doi.org/10.1126/science.1249097
15. Yang Y, Yang X, Liang L, et al (2019) Large-area graphene-nanomesh/carbon-nanotube hybrid membranes for ionic and molecular nanofiltration. Science (80-) 364:1057–1062
16. Ahmed, Z., Rehman, F., Ali, U., et al.: Recent Advances in MXene-based Separation Membranes. ChemBioEng Rev **8**, 110–120 (2021). https://doi.org/10.1002/cben.202000026
17. Wang, S., Yang, L., He, G., et al.: Two-dimensional nanochannel membranes for molecular and ionic separations. Chem Soc Rev **49**, 1071–1089 (2020). https://doi.org/10.1039/c9cs00751b
18. Li, J., Li, T., Ma, X., et al.: Regulating the Interlayer Spacing of 2D Lamellar Polymeric Membranes via Molecular Engineering of 2D Nanosheets. Macromolecules **54**, 4423–4431 (2021). https://doi.org/10.1021/acs.macromol.1c00056
19. Kang, Y., Xia, Y., Wang, H., Zhang, X.: 2D Laminar Membranes for Selective Water and Ion Transport. Adv Funct Mater **29**, 1–17 (2019). https://doi.org/10.1002/adfm.201902014
20. Kunimatsu, M., Nakagawa, K., Yoshioka, T., et al.: Design of niobate nanosheet-graphene oxide composite nanofiltration membranes with improved permeability. J Memb Sci **595**, 117598 (2020). https://doi.org/10.1016/j.memsci.2019.117598
21. Chen, C., Qin, S., Liu, D., et al.: Ultrafast, Stable Ionic and Molecular Sieving through Functionalized Boron Nitride Membranes. ACS Appl Mater Interfaces **11**, 30430–30436 (2019). https://doi.org/10.1021/acsami.9b08296
22. Thommes, M., Kaneko, K., Neimark, A.V., et al.: Physisorption of gases, with special reference to the evaluation of surface area and pore size distribution (IUPAC Technical Report). Pure Appl Chem **87**, 1051–1069 (2015). https://doi.org/10.1515/pac-2014-1117

23. Roth, W.J., Nachtigall, P., Morris, R.E., Cejka, J.: Two-dimensional zeolites: current status and perspectives. Chem Rev **114**, 4807–4837 (2014)
24. Khayum, M.A., Kandambeth, S., Mitra, S., et al.: Chemically delaminated free-standing ultrathin covalent organic nanosheets. Angew Chemie **128**, 15833–15837 (2016)
25. Yang, S., Zhang, K.: Few-layers MoS2 nanosheets modified thin film composite nanofiltration membranes with improved separation performance. J Memb Sci **595**, 117526 (2020)
26. Khulbe, K.C., Matsuura, T.: Removal of heavy metals and pollutants by membrane adsorption techniques. Appl Water Sci **8**, 19 (2018). https://doi.org/10.1007/s13201-018-0661-6
27. Hyun, T., Jeong, J., Chae, A., et al.: 2D-enabled membranes: materials and beyond. BMC Chem Eng **1**, 1–26 (2019). https://doi.org/10.1186/s42480-019-0012-x
28. Alsharaeh, E., Ahmed, F., Aldawsari, Y., et al.: Novel synthesis of holey reduced graphene oxide (HRGO) by microwave irradiation method for anode in lithium-ion batteries. Sci Rep **6**, 29854 (2016)
29. Tao, Y., Sui, Z.-Y., Han, B.-H.: Advanced porous graphene materials: from in-plane pore generation to energy storage applications. J Mater Chem A **8**, 6125–6143 (2020)
30. Meyer, J.C., Chuvilin, A., Algara-siller, G., et al.: Selective Sputtering and Atomic Resolution Imaging of Atomically Thin Boron Nitride Membranes. Nano Lett **9**, 2683–2689 (2009). https://doi.org/10.1021/nl9011497
31. Yi, M., Shen, Z.: A review on mechanical exfoliation for the scalable production of graphene. J Mater Chem A **3**, 11700–11715 (2015). https://doi.org/10.1039/c5ta00252d
32. Li, L.H., Glushenkov, A.M., Hait, S.K., et al.: High-efficient production of boron nitride nanosheets via an optimized ball milling process for lubrication in oil. Sci Rep **4**, 7288 (2014)
33. Xiong, J., Di, J., Xia, J., et al.: Surface defect engineering in 2D nanomaterials for photocatalysis. Adv Funct Mater **28**, 1801983 (2018)
34. Tan, C., Cao, X., Wu, X.-J., et al.: Recent advances in ultrathin two-dimensional nanomaterials. Chem Rev **117**, 6225–6331 (2017)
35. Jeon, M.Y., Kim, D., Kumar, P., et al.: Ultra-selective high-flux membranes from directly synthesized zeolite nanosheets. Nature **543**, 690–694 (2017)
36. Kandambeth S, Biswal BP, Chaudhari HD, Rout KC (2017) S. Kunjattu H, S. Mitra, S. Karak, A. Das, R. Mukherjee and UK Kharul. Adv Mater 29:1603945
37. Wang Q, Lei Y, Wang Y, et al (2020) Atomic-scale engineering of chemical-vapor-deposition-grown 2D transition metal dichalcogenides for electrocatalysis. Energy Environ Sci
38. Zhang, C., Wang, Z., Tu, R., et al.: Growth of self-aligned single-crystal vanadium carbide nanosheets with a controllable thickness on a unique staked metal substrate. Appl Surf Sci **499**, 143998 (2020)
39. Ji P, Chen J, Huang T, et al (2020) Fast preparation of vertical graphene nanosheets by helicon wave plasma chemical vapor deposition and its electrochemical performance. Diam Relat Mater 107958
40. Khan K, Tareen AK, Aslam M, et al (2020) Synthesis, properties and novel electrocatalytic applications of the 2D-borophene Xenes. Prog Solid State Chem 100283
41. Zhang, K., Feng, Y., Wang, F., et al.: Two dimensional hexagonal boron nitride (2D-hBN): synthesis, properties and applications. J Mater Chem C **5**, 11992–12022 (2017). https://doi.org/10.1039/C7TC04300G
42. Park, J.-H., Choi, S.H., Zhao, J., et al.: Thickness-controlled multilayer hexagonal boron nitride film prepared by plasma-enhanced chemical vapor deposition. Curr Appl Phys **16**, 1229–1235 (2016)
43. Hong J, Lee S, Seo J, et al (2015) A highly sensitive hydrogen sensor with gas selectivity using a PMMA membrane-coated Pd nanoparticle/single-layer graphene hybrid. ACS Appl Mater & interfaces 7:3554–3561
44. Huang, S., Dakhchoune, M., Luo, W., et al.: Single-layer graphene membranes by crack-free transfer for gas mixture separation. Nat Commun **9**, 1–11 (2018)
45. Ding, L., Wei, Y., Wang, Y., et al.: A two-dimensional lamellar membrane: MXene nanosheet stacks. Angew Chemie Int Ed **56**, 1825–1829 (2017)

46. Chen, C., Wang, J., Liu, D., et al.: Functionalized boron nitride membranes with ultrafast solvent transport performance for molecular separation. Nat Commun **9**, 1902 (2018). https://doi.org/10.1038/s41467-018-04294-6
47. Lin, H., Mehra, N., Li, Y., Zhu, J.: Graphite oxide/boron nitride hybrid membranes: The role of cross-plane laminar bonding for a durable membrane with large water flux and high rejection rate. J Memb Sci **593**, 117401 (2020). https://doi.org/10.1016/j.memsci.2019.117401
48. An, L., Bai, C., Bai, Y., et al.: Superhydrophobicity of polyvinylidene fluoride induced by integrating liquid-exfoliated hexagonal boron nitride nanosheets. High Perform Polym (2019). https://doi.org/10.1177/0954008319851173
49. Abdikheibari, S., Lei, W., Dumée, L.F., et al.: Novel thin film nanocomposite membranes decorated with few-layered boron nitride nanosheets for simultaneously enhanced water flux and organic fouling resistance. Appl Surf Sci **488**, 565–577 (2019). https://doi.org/10.1016/j.apsusc.2019.05.217
50. Shen, J., Liu, G., Huang, K., et al.: Subnanometer two-dimensional graphene oxide channels for ultrafast gas sieving. ACS Nano **10**, 3398–3409 (2016)
51. Yampolskii, Y.: Polymeric gas separation membranes. Macromolecules **45**, 3298–3311 (2012)
52. Nielsen, L.E.: Models for the permeability of filled polymer systems. J Macromol Sci **1**, 929–942 (1967)
53. Do Duong, D.: Adsorption analysis: equilibria and kinetics. Imperial College Press (1998)
54. Mi B (2014) Graphene oxide membranes for ionic and molecular sieving. Science (80-) 343:740–742. https://doi.org/10.1126/science.1250247
55. Sun, L., Ying, Y., Huang, H., et al.: Ultrafast molecule separation through layered WS2 nanosheet membranes. ACS Nano **8**, 6304–6311 (2014)
56. Zhang, J., Li, Z., Zhan, K., et al.: Two dimensional nanomaterial-based separation membranes. Electrophoresis (2019). https://doi.org/10.1002/elps.201800529
57. Liu, L., Ryu, S., Tomasik, M.R., et al.: Graphene oxidation: thickness-dependent etching and strong chemical doping. Nano Lett **8**, 1965–1970 (2008)
58. Boott, C.E., Nazemi, A., Manners, I.: Synthetic covalent and non-covalent 2D materials. Angew Chemie Int Ed **54**, 13876–13894 (2015)
59. Cheng, Y., Wang, X., Jia, C., et al.: Ultrathin mixed matrix membranes containing two-dimensional metal-organic framework nanosheets for efficient CO2/CH4 separation. J Memb Sci **539**, 213–223 (2017)
60. Sint, K., Wang, B., Král, P.: Selective ion passage through functionalized graphene nanopores. J Am Chem Soc **130**, 16448–16449 (2008)
61. Sun, C., Wen, B., Bai, B.: Application of nanoporous graphene membranes in natural gas processing: Molecular simulations of CH4/CO2, CH4/H2S and CH4/N2 separation. Chem Eng Sci **138**, 616–621 (2015)
62. Konatham, D., Yu, J., Ho, T.A., Striolo, A.: Simulation Insights for Graphene-Based Water Desalination Membranes. (2013). https://doi.org/10.1021/la4018695
63. Hafeez, A., Abdul, Z., Fauzi, A., et al.: Chemosphere Tuneable molecular selective boron nitride nanosheet ultrafiltration lamellar membrane for dye exclusion to remediate the environment. Chemosphere **303**, 135066 (2022). https://doi.org/10.1016/j.chemosphere.2022.135066
64. Ding, L., Wei, Y., Li, L., et al.: MXene molecular sieving membranes for highly efficient gas separation. Nat Commun **9**, 1–7 (2018)
65. Tsou, C.-H., An, Q.-F., Lo, S.-C., et al.: Effect of microstructure of graphene oxide fabricated through different self-assembly techniques on 1-butanol dehydration. J Memb Sci **477**, 93–100 (2015)
66. Peng Y, Li Y, Ban Y, et al (2014) Metal-organic framework nanosheets as building blocks for molecular sieving membranes. Science (80-) 346:1356–1359

67. Wang, S., Xie, Y., He, G., et al.: Graphene oxide membranes with heterogeneous nanodomains for efficient CO2 separations. Angew Chemie Int Ed **56**, 14246–14251 (2017)
68. Hung, W.-S., Tsou, C.-H., De Guzman, M., et al.: Cross-Linking with Diamine Monomers To Prepare Composite Graphene Oxide-Framework Membranes with Varying d -Spacing. Chem Mater **26**, 2983–2990 (2014). https://doi.org/10.1021/cm5007873
69. Carlsson, J.M., Hanke, F., Linic, S., Scheffler, M.: Two-step mechanism for low-temperature oxidation of vacancies in graphene. Phys Rev Lett **102**, 166104 (2009)

Polymeric Paints and Coatings

Zulfiqar Ahmad Rehan and Ahmad Usman

Abstract Polymeric coatings are applied on surfaces of metal, wood, plastics, and other materials to offer protection, decoration, and specific functionality. Polymeric coating technology is one of the old fields; still, it requires further maturity for perfection. The key trends in polymeric coatings include the production of environment friendly coatings, improving the functionality of existing coatings, and the development of smart coatings with multifunctional properties. Polymeric coatings with such properties are not feasible by conventional formulation and synthesis techniques. Therefore, modern technologies such as controllable graft polymerization, free-radical polymerization, and micro-emulsion polymerization are employed for these coatings. Moreover, the use of novel modified methods, functional pigments and the construction of nano- and micro-surfaces can produce polymeric coatings with multifunctional and enhanced properties. This chapter emphasizes traditional and advanced functional polymer-based coatings.

Keywords Polymeric coating · Latex paint · Elastomer · Water-borne coating · Varnish

1 Introduction

Paints and coatings are polymeric materials applied on numerous substrates through different techniques like dispersion, extrusion, and solution casting methods. The coatings based on polymers offer outstanding physical bonding while shielding from the environment. Polymer coatings are prepared by using components like resin, pigment, solvent, and additives. Each component imparts unique functionality in coatings, e.g., additives improve the functional properties, thinners or diluents lower the viscosity of the solution [1].

There is an ever-increasing demand for polymeric coatings especially in food industry, membranes, aerospace, marine, automobile, storage devices, oil and gas

Z. A. Rehan (✉) · A. Usman
Department of Materials, National Textile University, Faisalabad, Pakistan
e-mail: zarehan@ntu.edu.pk

K. Shaker and A. Hafeez (eds.), *Advanced Functional Polymers*, Engineering Materials,
https://doi.org/10.1007/978-981-99-0787-8_4

industry, etc. Some specific applications include protection against corrosion, breathable coatings in textile industry, encapsulation of electronic circuits, multilayer polymer coatings in infrastructure and transportation applications, and implantable devices [2].

1.1 Features of Polymeric Paint and Coating

In order to achieve gloss, reflectance, colour, flexibility, hardness/strength, and weathering resistance, different surfactants, pigments, fillers, binders and plasticizers are incorporated. The performance of polymer coatings significantly depends upon the interface [3]. When applied to a substrate, three distinct interfaces can be identified, namely interface with substrate at bottom, with air at top and with additives inside the coating. Further interfaces that correspondingly exhibit a role are gas–liquid, solid–liquid, solid–gas and interfaces (e.g., in drying paints) [4].

The chemistry of polymer is different as compared to that of the polymeric coatings. The fascinating characteristics of polymer coatings are that they are certainly spread and controlled. They can be employed as liquids in the case of paints. Certain additional characteristics such as drying, flexibility, and self-healing are the ones where molecular movement performs an essential role [5]. The innovative polymeric coatings are thermosetting in nature, regardless of the fact that majority of polymeric materials are thermoplastics [6].

2 Physically Drying Polymeric Coatings

Latex paint is a broad word for substantially drying paints based on aqueous polymer pigments, dispersions, and fillers. Latex paints are categorized by a Pigment Volume Concentration (PVC) content >40% and latex sheen enamels, through a PVC <40% (Table 1).

Table 1 Classification of latex paints and latex gloss enamels

Paint type	PVC [vol%]	Binder contents [wt%]
1. Latex gloss enamels		
Satin	30–40	38–33
Medium-gloss	15–30	65–38
Glossy	10–15	70–65
2. Latex paints (flat)		
Low-cost interior	80–90	10–5
Interior	60–80	21–10
Exterior (facades)	40–60	33–21

The adhesives are predominantly core dispersions (e.g., styrene-acrylate or pure acrylate). Acrylates of pure origin are ideal for weatherable shine enamels [7].

2.1 *Latex Paints*

Latex paints for indoor applications of low-cost have a Critical Pigment Volume Concentration (CPVC) < PVC (rarely CPVC $\ll$ PVC) and have an advantage to permeable finishes including a (appropriately) elevated flow of water vapor.

In comparison with other latex paints, superior value latex paints for façade possessed higher value of CPVC than PVC and that leads to high quality non-permeable varnishes with minimal absorption of water. Moreover, latex paints include substantial volumes of (low-cost) additives like titanium dioxide as white pigment, making latex paint more appealing aesthetically. Pigments of inorganic nature having refractive index number of less than 1.7 are mostly known as extenders (fillers), and those including refractive index number more than 1.7 are believed to be pigments of white nature [8]. If the PVC is more than the CPVC, air/filler interfaces are produced in the coating [9]. Therefore, the extenders (fillers) may reduce the strength of hiding over CPVC ("dry-hiding" effect).

2.1.1 Latex Gloss Coating

Latex glossy enamels produce coatings/paints through comparatively superior gloss in comparison to latex paints. Glossy latex enamels produce sheen by three factors:

- The pigment to volume concentration would remain as minimal as possible, and it should be less than 20% for glossy gloss latex enamels.
- The polymer dispersion usually acrylate (pure) must have particle size of 0.1 μm or less.
- Suitable modifier incorporation in latex paints for improved rheological behavior.

Additionally, polymer dispersants are incorporated after grinding the pigment during the production of simple latex and glossy paints in order to avoid agglomeration during pigmentation [10].

2.2 *2 K Water-Borne Polyurethanes (PU) Coatings*

Poly-isocyanates may possibly react with water in case of water-borne two-components PU paints. Primarily, an unbalanced intermediate mixture, an alternate carbaminic acid, has being produced. Therefore, instantly converted to CO_2 (carbon dioxide) as well as amine (primary). This primary amine develop urea linkages

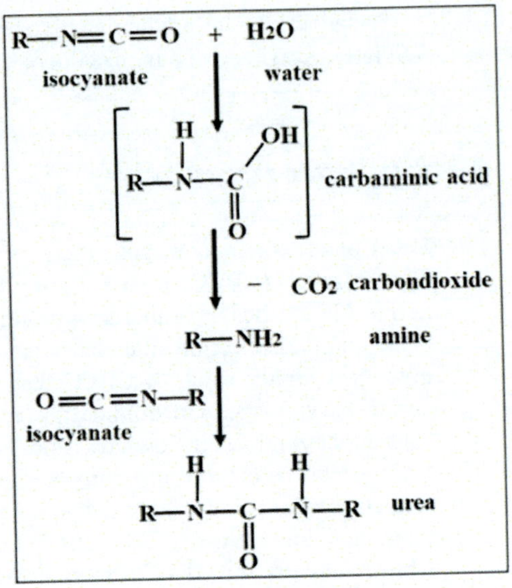

Fig. 1 Reaction of isocyanates with water

through instinctively combination with other isocyanate groups [11]. This reaction kinetics are shown in Fig. 1.

The reaction preferably takes place at the amine groups or hydroxyl groups of resins. The research display that there are likely chances of any reaction happened with the carboxylic groups of water-diluent resins. In order to get the reaction kinetics of these compounds, the two components be dissolved effectively and has significant impact [12]. Usually by the addition of polyisocyanate as hardener to water-borne systems, coarse emulsions are produced. The ingredients are grinded to produce small particles by diffusion methods. That's why, the possibility of an effective crosslinking reaction between the resins and poly-isocyanates exists.

2.2.1 Reaction Kinetics for Poly-isocyanates in Water-Borne Coatings/Paints

The transfer method of hydroxyl-based resins in 2-component PU (polyurethane) processes should never include hydroxyl groups and solvents [13]. Triethyl amines (e.g., Tertiary amine) are ideal that never encompass further active groups, e.g., NH groups or hydroxyl groups, for neutralization. Though the effective homogenization based on the selection of resins and hardeners, its intended use and reactions among isocyanate and water should be taken into account.

2.2.2 Application Area of 2 K-Component Water-Based Polyurethane Paints

Similar solvent-borne systems, ready to use water-borne polyurethane blended paints have a limited pot-life [14]. They are prominent for tremendous chemical resistance and weather resistance if aliphatic or cycloaliphatic polyisocyanate adducts are selected. Releases of organic compounds (VOC) are substantially lower than in the case of solvent-borne products. Consequently, application fields are those involving high-performance coatings and low emissions [15].

2.3 2 K Water-Borne Epoxy Coatings

In water-borne 2 K-components epoxy paints, hardener (polyamines) as well as the base component (epoxy resin) can be pigmented and diluted with water. The fillers and pigments must have the shortest-potential matter of soluble salts (electrolytes) for all water-borne paint methodologies. The solid-epoxy resins are dispersed in water, termed as emulsifiable epoxy-solid resin are mostly used as liquid coatings. It is already understood that highly viscous resins are difficult to emulsify, so the viscosity must be regulated in order to get better emulsification. Moreover, the performance of epoxy resin emulsion having medium viscosity is much better as compared to high or low viscosity resins. The characteristics of epoxy coatings and epoxy resin emulsions are mostly dependent of particle size of emulsion [16]. The course or rough emulsions have limited performing parameters are compared to fine particle size epoxy emulsions (Fig. 2).

Phase inversion is the best method for the emulsification of epoxy-based resins (Fig. 3). Initially, emulsification of epoxy resin (w/o emulsion) was performed and in

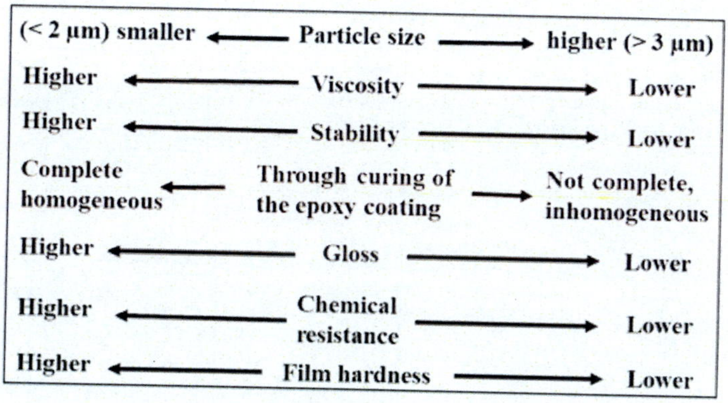

Fig. 2 Impact of the particle size on the properties of epoxy emulsions and coatings

later stage further addition of water through phase inversion produce the true epoxy resin emulsion in water (o/w emulsion).

It is observed that viscosity goes higher until at maximum value during the phase inversion (Fig. 4). At this moment, the kinetics can be studied in such a way that the particle size of oil/water emulsions can be affected by changing the shearing time and shear force [17].

It is observed that amine hardener doesn't crosslink homogenously with epoxy resin emulsion and coatings. Later on, these two reactants combine or crosslink in such a way that hardeners molecules exist in the epoxy resin emulsion and coatings (Fig. 5). Therefore, the hardeners molecules diffuse into the epoxy resin particles [18]. Hence, simply more than 70% of the reactive groups take part in reaction with each other. So, it can be experimentally stated in case of 2 K-waterborne epoxy coating and paint systems the hardeners are less consumed stoichiometrically.

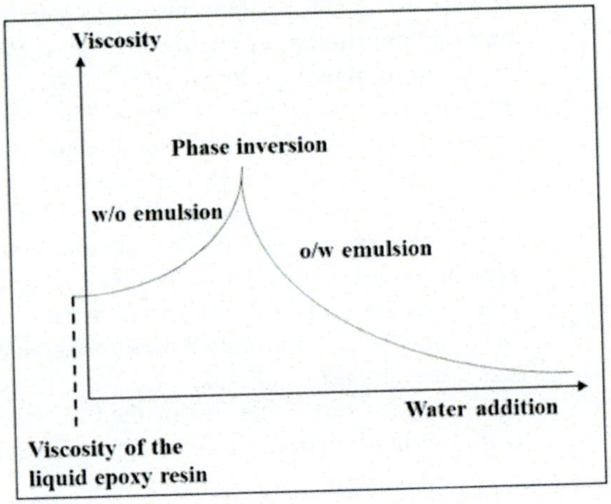

Fig. 3 Basic illustration of the difference in viscosity through emulsification of liquid epoxy resins by phase inversion

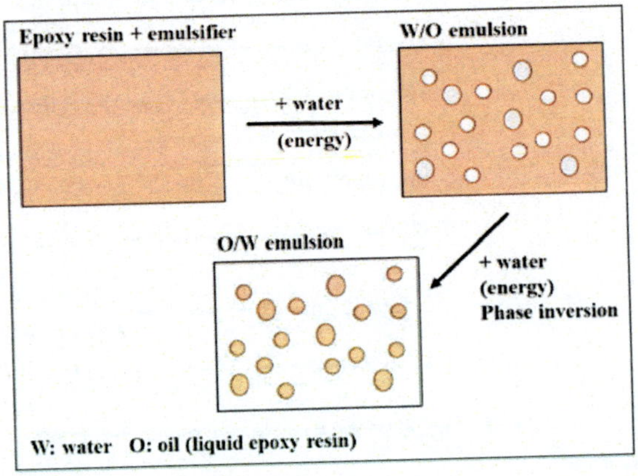

Fig. 4 Emulsification process of liquid epoxy resins by phase inversion

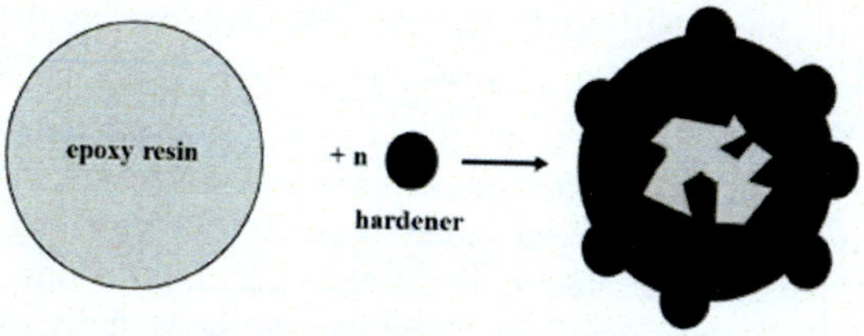

Fig. 5 Basic diagram of the reaction among a single epoxy resin emulsion droplet and amine hardeners

2.4 *Water-Borne Stoving Enamels*

Stoving enamels are categorized as solvent-borne systems, comprising of two components i.e., cross linkers and hydroxyl resins. Most importantly, these two components should be blended homogenously. All the reaction components should be blended uniformly by diffusion method in water borne techniques due to the colloidal nature as compared to organic coatings/emulsions [19].

2.4.1 Water-Borne Stoving Enamels Based on Amino Resins

Amino resins through non-etherification and etherification with methanol are water soluble in all the case. The studies revealed that resin produced through non-etherification method need special storage facility due to the non-compatibility with other resins. On the other hand, resin produced by methanol etherification are compatible with water-borne paints. The most prominent and well established amino comes under the category of melamine resin. The imino functionalized resin through etherification are relatively more soluble in aqueous media as compared to highly methylolated, fully etherified types (HMMM resins). Though, these methylolated resins are stabilized with anionically stabilized resin or with some general co-solvent [20]. Further the hydrolysis of these imino functionalized resin is resistant to both acidic and basic atmosphere. The mechanism of reaction is shown in Fig. 6 [9].

The alkyd resin are extremely suitable cross-linkers in melamine-based water borne paints. water-borne paints based on alkyd resin consequently give optimal wetting of substrate coats and pigments, the film fabricated through these resins have excellent gloss and durability. The melamine resins are now being used as primer, single coat paint as well water borne upper coat paint [21].

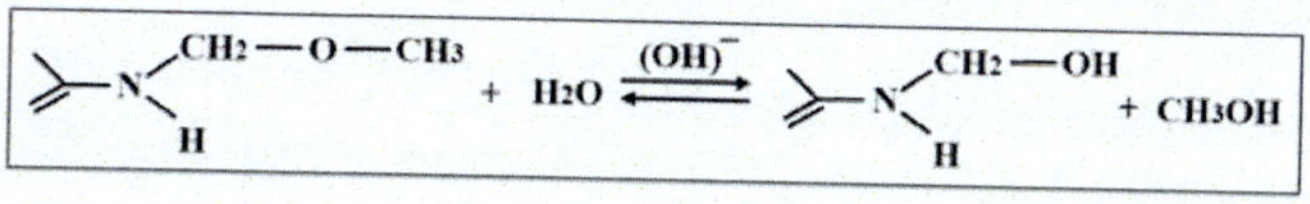

Fig. 6 Hydrolysis of amino resins

3 Solvents-Borne Polymeric Coatings

"Natural drying solvent-borne paint" is a term used to describe paint that is made by dispersing polymeric binder in a solvent-based system and drying naturally over time. The higher molecular mass of the polymer utilized as a binder in paints produces the unique qualities in the paint film. Furthermore, the structural properties of polymers are responsible for maintaining the evaporation rate during paint film formation. One of the drawback of using these polymeric binder is that a huge amount of solvent is use for smooth paint application [22].

Different type of binder is used which includes polystyrene, cellulose nitrate and their co-polymer, some halides and polyester of vinyl-based polymers, derivate of rubber, epoxy and hydrocarbon of higher molecular weight as well as polyamide and polycarbonate-based resin.

3.1 CN (Cellulose Nitrate) Based Polymeric Coating

Naturally occurring cellulose is the main component of all existing plants. It is based on carbohydrate of higher molecular weight comprising of glucose repeating units associated by β-1,4-glycoside linkages. They have higher level of boning as well relatively higher molecular weight molecules (molar mass about 106 g/mol) as shown in above Fig. 7. Cellulose nitrate (CN) was indeed more prominent in polymer class [23]. This polymer is produced through the reaction of naturally produced cellulose with other reactants like water, nitric acid and sometime sulfuric acid as well. The resultant polymer (cellulose nitrate), produced as finished product is similar to wool yarn extensively known as "nitrocotton". Later on, this product is soaked in different alcohols. Consequently, the chips of cellulose nitrate is produced by reacting plasticizer (phthalate esters) with cellulose nitrate (CN) [24].

The different grades of cellulose nitrate vary with different characteristics like molecular weight and the content of nitrate ester in resultant polymer. One of the impacts of molecular weight is the rise in viscosity of resultant solution. Classification with a high ester content are readily soluble in glycol ether esters, ketones, esters, off course be blended with aromatic hydrocarbons and alcohols [25]. The Chemical structure is demonstrated in Fig. 8.

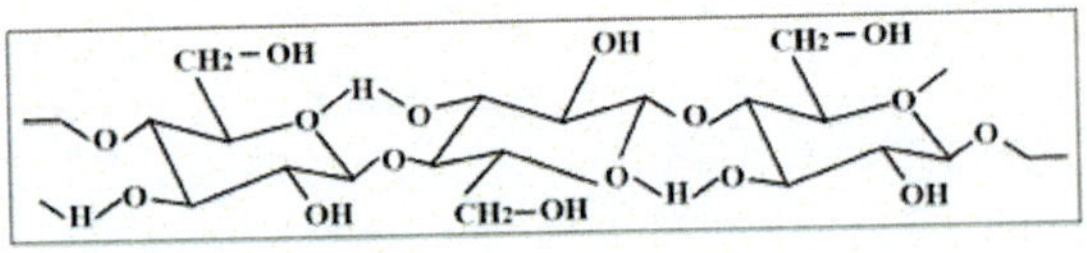

Fig. 7 Molecular formation of cellulose (ß-1,4-glycoside)

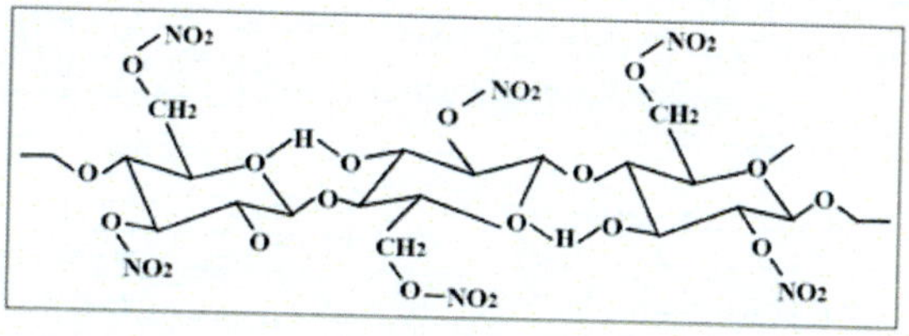

Fig. 8 Molecular formation of an ester-soluble cellulose nitrate

3.1.1 Blend Associates for Cellulose Nitrate

Cellulose nitrate is blended with another particular polyesters established on alkyd resins, and sebacic acid or adipic acid. The mono-carboxylic acids and short oil grades are fatty acids, known as alkyd resin comprise considerable quantities of synthetic acid like oleic, linoleic acid (e.g. based on peanut oil) [26]. Special amino resins (binders) are also appropriate for compatibility with cellulose nitrate. These resins (alkyd) are mostly used to produce pastes of pigment through pigmentation in order to get resultant cellulose nitrate-based paints.

3.1.2 Applications of Cellulose Nitrate Paints

The most well-known, higher molecular weight cellulose nitrate (CN) varnishes are good protectors when applied to metallic surfaces like copper, brass, and silver. These coatings produce clear gloss top layer giving aesthetic feel as well. These cellulose nitrate (CN) paints have also excellent naturally air-drying feature making it more prominent in the field of paints and coatings specially in crackle finishes and hammer finishes. This feature is more visible in printing coatings, topcoats and primers of naturally air-drying wood-varnishes, paper coatings and leather finishes as well [27].

3.2 Elastomeric (Modified Rubbers) Coating

The main component of rubber made from natural sources make up about 35% of its total weight. This type of rubber is produced by unique Euphorbiaceous trees. The chemical name is poly-2-methylbut-2-ene (polyisoprene) and the structure is shown in Fig. 9. The $-CH_3$ group is coordinated to the hydrogen atom in a cis-position on the isoprene portion. This is of great interest that molecular weight of this polymer could be about 10^6 g/mol. Now, in addition, efforts have been put together by the researchers to provide the appropriate naturally produced for paints and coating applications [28, 29]. There had been great effort in this field and two recently produced naturally rubbers, cyclized rubber (RUI) and chlorinated rubber (RUC) are used as binder in elastomeric paints and coatings applications.

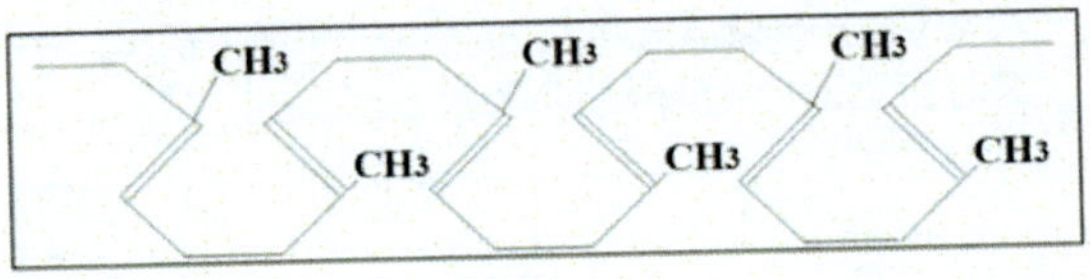

Fig. 9 Structure of rubber (poly-2-methylbut-2-ene) polyisoprene

3.2.1 Cyclized Rubber

This class of rubber is produced through cyclization of rubber at much higher temperature in the presence of catalyst (initiator) usually strong acids. The isoprene units form six-membered ring systems, and the large molecules decay to molar masses of 3,000–10,000 g/mol. Cyclized rubber is a stiff, resin-like material, and melts at temperatures among 120 and 140 °C. It is soluble in aromatic hydrocarbons and aliphatic and in terpenes, thinnable with ketones and esters, but insoluble in glycol ethers and alcohols [30]. For paint formulations, cyclized rubber has to be blended with a plasticizer. The most widely used brand of this class is Syntotex RR 800 (DSM).

3.2.2 Chlorinated Rubber

This class of rubber is prepared by first dissolving natural rubber in solvent carbon tetrachloride (Ccl_4) and then reacting with chlorine (cl_2) to produce chlorinated rubber. The substitution of chlorine occurs while addition and substitution after breaking down the cyclic structure of macromolecule. The chlorine contents are about 67 wt% in the resultant finished product having approximate molar weight from 30 to 180 KD (kilodalton). The composition is mentioned in Fig. 10. Currently chlorinated rubber is now derived from synthetic poly-isoprene and now not produced from natural rubber. This rubber is extremely soluble in ketones, esters and aromatic hydrocarbons and the appearance is usually white powder [31]. Though the product produces films merely by naturally air drying, ultimately excellent resistant to chemical. The film films produced through this class of rubber has excellent adhesion properties with substrate. Some of the types of this rubber having low molecular weight don't need plasticizer for elastic behavior of film. This elastic behavior provides excellent properties in special type of paints. Commercially, this typical rubber is economical product for some of the specialized paints. Few of the commercial products from Bayer (Germany) are Pergut S grades (5, 10, 20, 40, 90).

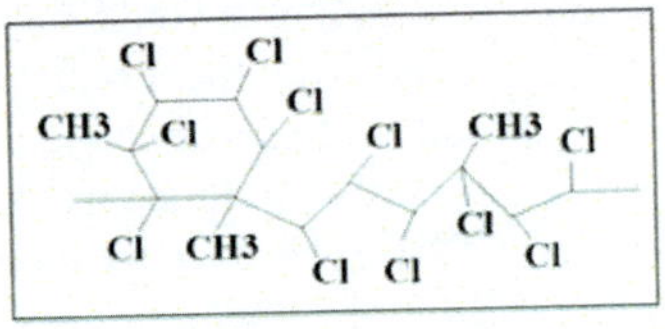

Fig. 10 Molecular formation of a segment of chlorinated rubber

3.2.3 Modified Rubber Applications

In order to produces functional paints against corrosion, rubber is modified with some other active agents. These functional paints based on modified chlorinated rubber are used in marine as well as industrial applications. One of the advantages of these cyclized modified rubber is the uniform permeation inside the concrete and asphalt substrates. Therefore, due to the excellent property of alkali and acid resistance, such functional pains have best application in indoor and outdoor swimming pools and other allied areas. Recently, the different manufactures are utilizing this modified chlorinated rubber into road-marking paint applications.

3.3 *Oxidatively Cure Paint*

The term air drying, or oxidative drying happens since the previous study reveals about the formation of solid film through transesterification of different vegetable oils and dibasic acids, later on this film is dried in natural air [32]. The chemical kinetic is not simple as the reaction proceeds in two step process and final solid material of characteristics properties are produced through condensation polymerization. The advantage of this functional paint produced through this phenomenon is drying of film formed in natural air/atmosphere.

3.3.1 Alkyd Resins

Alkyd or oil based-resins are however the extremely significant category of surface coating adhesive, used in enamel paints. This class of adhesives have commercially available with different grades depending upon their oil length and other components. These resins are produced through transesterification technique. Firstly, intermediate resin is formed through Mono-esterification step by the reaction of polyhydric alcohol (glycerol/glycerin, pentaerythritol and trimethylolpropane) and vegetable/fatty acids. In second step, monoesters are reacted again with dibasic acids like phthalic anhydride, maleic anhydride to produce alkyd resin of specific molecular weight through transesterification step via condensation polymerization [33]. The various classification for naturally drying paints having low –OH contents ($\leq$50 mg KOH/g), and high molecular weight, and low acidic value ($\approx$3–20 mg KOH/g). The classification of alkyd resins is categorized three (03) subclasses depending upon the type and amount of oil/fatty acids.

- Medium oil alkyds 40–60%
- Short oil alkyds Below 40%
- Long oil alkyds More than 60% oil or fatty acid

One subclassification is the long oil and is mainly produced by the condensation polymerization of polyhydric alcohol and pentaerythritol. The chemical kinetics

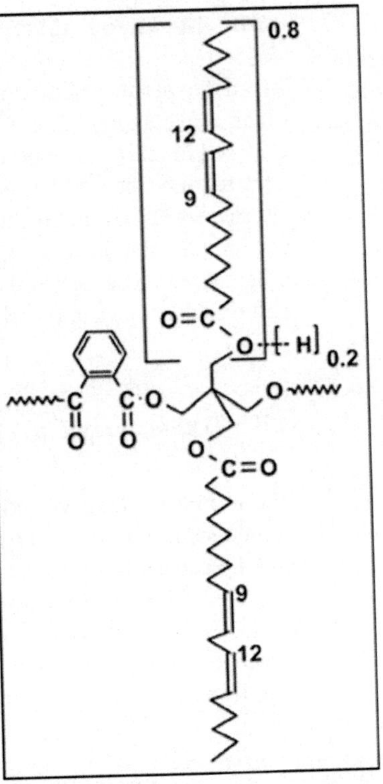

Fig. 11 Fundamental chain of a long oil alkyd resin

depends on the mole ratio of all the reactants. 1.0 (one) mole of polyhydric alcohol i.e., pentaerythritol and 1.8 part of linoleic acid (see Fig. 11), 1 mol phthalic acid, comprises 60.0 wt% fatty acid, 16.8 wt% pentaerythritol and 19.0 wt% phthalic anhydride.

3.4 *Alkyd Based Paint*

One of the most widely used enamel paints (solvent-borne) is based on alkyd resin.

Alkyd Resin (Long Oil), these kinds of special resin used in decorative enamel paints are manufactured by various kinds of edible oils like sunflower, linseed oil, soyabean oil safflower oil and cotton seed oil as well as the main raw material. These are also termed as fatty acids resins. In addition to decorative domestic paints these resins are utilized for corrosion protection paints. **Medium oil-based resin**, these resins are also prepared from edible oils and fatty acids as well. The amount of oil used in the process of manufacturing resin is less than the contents used in long oil resins. The enamel paint prepared from medium oil resin has the advantage of non-yellowing feature making it best suitable in the area of applications requiring fast-drying of paints, e.g., paints for radiator, agriculture equipment, industrial structures and automotive applications. The third category of resin produced from:

Modified urethane resin, modified urethane is produced for such applications where fast drying is required. They are best applicable for floor varnishes, printing inks and all the wood series coatings and coating including sand sealers [34].

3.5 2 K Polymeric Coatings (PU, Epoxies)

All chemically cured coating techniques have two basic mechanisms. This is possible only in the case where the two polymeric components are blended physically before application comply with the description of a two-component system [35]. The best-known good example is 2 K polyurethane component paints and 2 K epoxy coatings and paints. These coatings are termed as two components-based systems; therefore, it is well understood that one of the components presents in large amount i.e., comprised of pigment is known as base component. Similarly, the other component consists of small amount and known as hardener. The hardener is usually without pigment.

3.6 Two-Components Polyurethane Paints

Reactions of isocyanates: At ambient temperature, coordinated complexes of alcohols, primary and secondary amines with adjacent hydrogen react with isocyanates. Physically blending of isocyanate and hardener with reactive functional groups is taking place here. Commercially, the paint produced by the cross-linking reaction of the urethane group are known as 2 K-polyurethane paints (see Fig. 12). The process can be accelerated towards right side by addition of initiator or catalyst as the reaction kinetics of resin having hydroxyl as functional group with isocyanates proceeds at room temperature.

The film produced through this cross-linking of oligomers of HDI, and hydroxyl functionalized resins are not only had weather resistant properties but also the chemical resistant characteristics. Based on these key characteristics, the generated paints

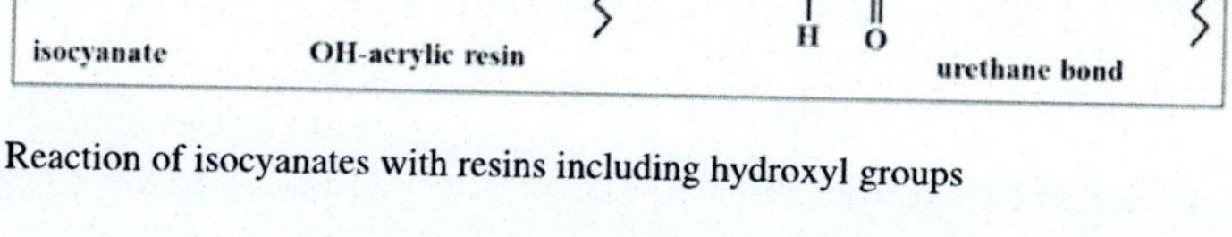

Fig. 12 Reaction of isocyanates with resins including hydroxyl groups

and coatings are best suited for specific applications including automotive painting of vehicles, agriculture equipment, locomotive, repair paints and OEM coatings [36].

3.7 Two-Components Epoxy Paints

All types of paints and coatings produced from epoxy resins must have some excellent chemical resistance. Additionally, they exhibit resistance to abrasion, adhesion, and film hardness of the produced paint on various metallic substrates. One of the excellent features of 2 K paints based on epoxy resin is the protection against metal corrosion and protection in industrial application. The epoxy resin is produced with different ingredients and one of the types of epoxies produced from aromatic bisphenol A resin have comparatively low weatherability. The mechanism of curing of epoxy-based paint is entirely different as the curing mechanism of urethane based occurs by polyaddition of curing agent and hardener at ambient temperature [37].

Base component: Bisphenol A is one of the components used in the production of epoxy resin. This again synthesized by complex reaction of 1.0 mol of acetone and 2.0 mol of phenol. The epoxy, bisphenol A-diglycidyl ether is produced by reaction of 2 mol of ECH (epichlorohydrin) and 1.0 mol of bisphenol A. The epoxy resin can be produced of different molecular weights and such different kind of epoxies can be produced by changing the stoichiometric or mole ratio of these reactants and reaction parameters (Fig. 13).

In normal practices for two-components techniques, the epoxy resins (major ingredient) and hardener (amino functional) are physically blended with each other before applying to some specific area of application. The curing of material takes place after blending the epoxy resin having oxirane group and the later on addition of cross-linker i.e., hardener in 2-K method of paints and coatings. The detail kinetic behavior is shown in Fig. 14. As it is very much clear that the reaction takes place in two steps and in the case of epoxy-based paints and coatings, the first step of the reaction proceeds faster as compared to the second step due to the fact that the primary amines have less steric hinderance as compared to the secondary amine.

The hardeners based on amino functionalization is characterized by the equivalent weight of –NH group. The amino functional molar weight is defined as the molar

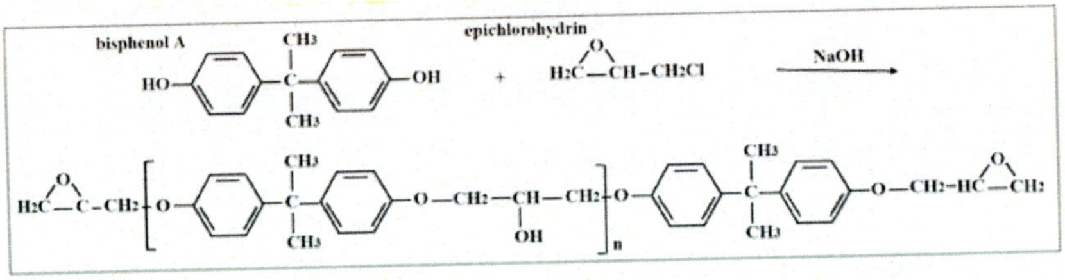

Fig. 13 Synthesis of epoxy resins based on bisphenol A ($n = 0$: bisphenol A—diglycidyl ether)

Fig. 14 Curing mechanism of epoxy-based resins

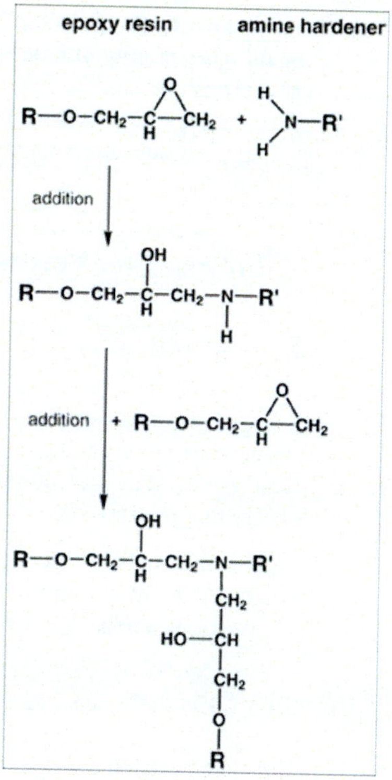

weight of –NH hardener by dividing it with active hydrogen atoms on –NH (amino) group and ultimately termed as 1 mol of active hydrogen [38].

4 Stoving Enamels

The class of paint mostly used in industrial application are termed as stoving paints. They are cured at comparatively high temperature but are best suitable in industrial coatings. For the reason that of the numerous types of techniques in reality, useful baking/stoving parameters. The most important is the temperature and the range of stoving temperature is around 70 and 240 °C. Usually the Stoving/baking cycles is as minimum as one minute and as long as one hour as well. Under these situations—for the duration of film formation—crosslinking takes place.

The mechanism of stoving is very interesting, one or two types of alkyd resins are utilized to react in such a chemical manner that final molecular structure is formed by strong cross-linking with each other. It is also important to mention here that some of the mechanism for stoving paints are through self-crosslinking [39]. The study shows that some of the Stoving enamels are also classified into reactive stoving paints. The stability of paint is one of the saliant feature in stoving paint and hence the components and ingredients of resins are selected in such a way that the final paint remains in useable condition for longer time without any initiation towards gelling at storage temperature [40].

5 Solvent Free Polymeric Varnishes

5.1 *Solvent-Free 2 K Epoxy Varnishes*

Some of the coatings are produced, based on epoxy resin and are cured at room temperature without incorporation any solvent. The cross-linking mechanism is same as found in solvent-based methods. There are two most important variations concerning preparation:

- Low-molecular weight epoxy resins based on liquid epoxy can be produced.
- The different solvents or thinners are added to reduce the viscosity of the final coating solution.

5.1.1 Epoxy Based Liquid Resins

It is already stated above that the epoxy either solid or liquid are based on two major components. One of the components is bis-phenol A and other is epichlorohydrin and the equivalent weight of epoxy for liquid should be greater than 180 as depicted in Fig. 15. The cost and performance depend upon the equivalent weight of epoxy known as di-glycidyl ether epoxy.

The liquid epoxy resin is available commercially in different grades depending upon the molecular weight and the epoxy having low molecular weight have comparatively less shelf life as well as lower drying period. On the other hand, the characteristic features of solid epoxy related to quick drying usually termed as touch to dry phenomenon. This feature is most important due to the use of solid epoxy in certain application like powder coatings.

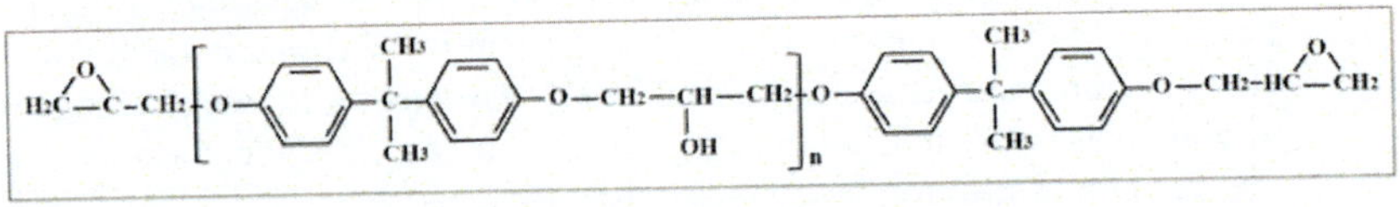

Fig. 15 Bisphenol A-epoxy based resins

5.2 *Radiation Curing Systems*

When certain complexes, primarily those with double conjugated bonds, are exposed to very energetic radiations, applications requiring excellent crosslinking are chosen. For some applications, it is possible to chemically cross link low molecular weight components with higher molecular weight component within the matrix. But the reaction kinetics are highly exothermic and efficient as well. In most cases, low molecular ingredients are low viscous liquid that can be used in solvent-free radiation cure systems recipes as volatile liquids [41]. This system is further subdivided into other two subcategories. One is cured by ultraviolet light (UV) and other through electron beam (EBC).

5.3 *UV Curing*

UV drying appears as paint film comprising unsaturated complexes are cured with UV radiation. There is special UV radiation generator to produce this light of specific wavelength. This specific wavelength radiation react special initiator (UV initiator) also termed as photo initiator to produce free-radicals. These radicals initiate the radical polymerization of conjugated unsaturated complexes. Occasionally, the most important is the blending of photo-initiator with photo-sensitizers. The cross-link or bulk density of coating If the conjugated blend of the coating is dependent on the conjugated crosslinking due to unsaturation of individual component. The cation on later stage could be produced by initially catalyzed reaction through ultra-violet radiation with epoxy resin, facilitating further cross-linking. Now a days this is most favorable reaction process in UV-based coatings and paints. High-voltage electron beams can produce free-radicals in unsaturated complexes immediately. There is another innovation in this field to produces free-radicals by field emission gun or tungsten lamp to produce high energy electron. The advantage of this process is that the photo initiator is not required to build a molecular network through free radical polymerization's cross-linking.

5.3.1 *Mechanism of UV Crosslinking/Curing*

In order to understand the mechanism of UV crosslinking or curing, we must know that the ultra-violet (UV) light is type of electromagnetic radiation that have wavelength in the range of 100–400 nm. The smaller the wavelength, the higher the energy of these radiations. The visible light has little bit higher wavelength. The energy of the whole spectrum of radiation is different as it is related to wavelength. The energy in of UV-radiation is from 300 to 760 kJ/mol. Such energy of UV light can fall on the molecular structure and stimulate the conjugated double bond resulting in initiation of chemical reaction between the chemical components. This is now interesting to

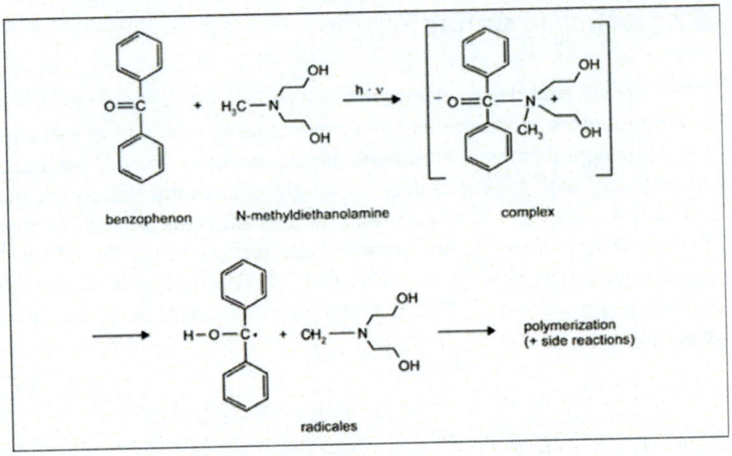

Fig. 16 Formation of free-radical from benzophenone

understand that these high energy in this wavelength region (100–280 nm) cure the paints and coating film initially at the surface via this UV-curing technique. After the partial curing of film at the surface of coating, the complete curing can be performed in visible wavelength region (315–380 nm) [42].

5.3.2 UV Sensitizers and Initiators

The phenomenon of free radical formation is already explained where, UV radiation split the phot-initiator into free-radicals via double bond breaking. One of the interesting features of photo-sensitizer is the absorption of ultra-violet radiation and then fall this absorb energy to other reacting species or molecule to free-radical via intermolecular reaction mechanism [43, 44]. Benzophenone is a well-known ultra-violet initiator. The generation of free-radical through the alky derivative also leads to hydrogen as donor (Fig. 16).

5.3.3 Resins for UV Varnishes

Saturated and unsaturated resins can be used for all kind of coatings but the best ultra-violet curing application, unsaturated resins are the best suitable. These resins can be produced by the polycondensation reaction of maleic and phthalic anhydride with glycols. The unsaturated resin is marketed in diluted form and the best suitable solvent for this resin is styrene. Such class of resins applied in the final application through ultra-violet curing technique need to be modified in order to prevent from oxidation during drying method. (e.g., Isophthalic anhydride, ether modified polyols). Resins with acrylic groups are the following, further significant class of special resins to cure the coating through ultra-violet mechanism. So broadly speaking, unsaturated modified acrylic resins are best applicable actually either polymer or oligomers for this curing methods. Pure acrylic adhesives or resins are produced through the reaction of methacrylic acid or methacrylic esters and the other approach is to produce from acrylic ester or acids. The best suitable for the above-mentioned curing method is through acrylic ester and acids. The unique feature can be incorporated through integrating resin molecular structure with acrylic acid derivatives. The subsequent cross-linked resin occur: polyester acrylates, acrylates based epoxy, acrylates based polyurethane, unsaturated acrylic resins, polyether acrylates, and silicone acrylates [45].

5.3.4 Characteristics and Application of UV Coatings

Coatings based on ultra-violet have astonishing features about the scratch resilience and impact resistance. There is little hard film formed by this technique. In order to develop elasticity or softness in the film, different plasticizers are added in the material that plasticize the structural arrangements of the molecules of resins and also the thinners give addition elastic response to film as well [46]. These additional significant attributes make it best suitable for lamination sheets, wood polish, home and office furniture and sealers as well. The other area of application refers paper coating, foil coating. Similarly, printing inks, laminates, printing sealers can also be prepared by ultra-curing techniques.

5.4 Curing by Electro-Beam

This technique is one of the latest techniques and being used in special applications. The high energy electron is the main source of curing the coating film. These electrons are generated from electron generation source, one of them could be tungsten lamp and other may be field emission gun. The free-radical are created by accelerated voltages from 120 to 500 kV. The active electrons from molecular bonding and the accelerated electron from field emission source react to form cationic configuration though breaking of chemical bond and similarly anionic formation occurs by the addition of electron. So, this phenomenon produces free-radicals from these cation and anions. These accelerated electrons in some condition breakdown the double bond and produce more free radicals. The more the energy of these electron, the more will be free radical formation as well. The energy of these electrons coming from electron source is directly related to the applied voltage to the source. This very well established technique to cure the paints and coating film by electron bombardment [47].

This technique has the benefit of not requiring an initiator for any form of treatment. As we have observed in previous method that there is always a cross-lining phenomenon required for film curing but, in this technique, no need of any cross-linking at the top, or in between two layers of coatings with pigments. As the accelerated voltage of electron beam is higher as compared with ultra-violet radiation, therefore a fast movement of conveyer belt is required for quick action to cure. Unlike the other curing technique, this technique is much better and have excellent surface adhesion properties with plastic substrate. The mechanism behind the adhesion of coating on plastic substrate is the free-radical formation due to the high energy electrons. This reaction is termed as grafting technique [48]. There is only one limitation in this grafting technique i.e., the inert gas is usually required to accelerate the electron. This inert condition reduces the cross-linking efficiency as compared to oxygen or air which facilitate the curing step.

5.5 *Powder Coating*

In early 1960s, specialized spray gun was employed for powder coatings through the phenomenon of electrostatic deposition. Following this spray gun, the numerous fields were established using powder coating [49]. The markets still emerging in different area of applications. To till date the powder coating is the only solution to get rid of organic solvent spray coatings and as a result green solution for paints application. There are two different approaches for environmentally stable solution. Firstly, coating material from powder are virtually free from any solvent. Secondly, after coating powders can be reprocessed [50].

5.5.1 Manufacture of Powder Coating Materials and Universal Properties

We already studied that powder coating materials are free from any organic solvent; this significance encourages the industries to invest in this technology for bulk production of material. Consequently, the powder is made predominantly in melt methods. First of all, different ingredients are first blended and then pass-through hot melt. The viscosity is the main feature in these hot-melt condition. The more viscous materials were then brought out from these hot chambers or closed reaction towers. Then this highly viscous material is passed through series of dryers to convert this slurry to semisolid form. Later on, this semi dried material is passed though the twin extruders. The extruder converts the material into small portions followed by mixer to produce powder of big particle size. In case, the extruder is operating continuously, it is recommended the mix together all the ingredients prior to extrusion process. One precaution is to avoid using liquid components during the extrusion process [41]. Technically make sure that liquid contents must be incorporated prior to extrusion in hot-melt process. Later on, the master batch is prepared by converting this highly viscous semi solid material into small grains and then this granular mixture is added to another physical mixer to segregate the smaller particle

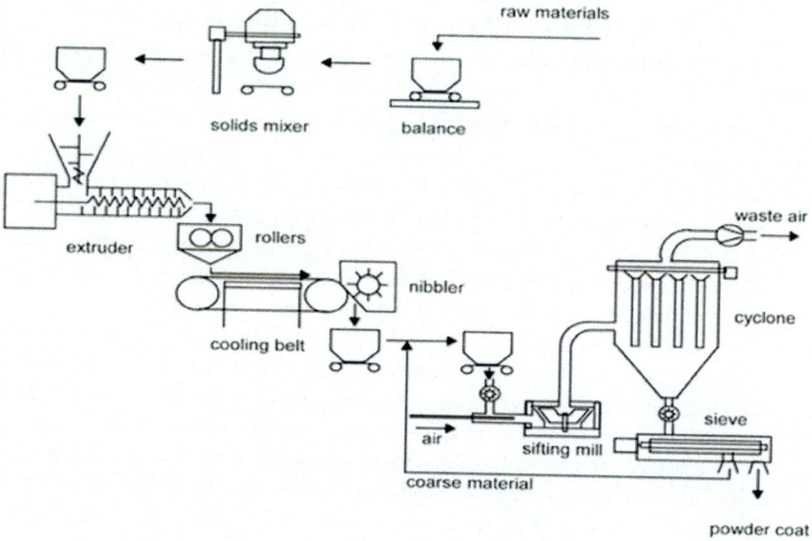

Fig. 17 Procedure for the production of powder coatings

from bigger fractions. In the almost last step, the pigments are also added along with this granular material and consequently, the final shape of the powder coating material is relatively brilliant appearance. The extruded material passes all the way from series of roller forming heavy material block shredded by shredders called nibblers. Special king of pestle and mortar are used to crush the material into fine powder as much possible as can be. The crushed product must be placed in cool place. The fine particle is extracted by special cyclone and the course particles are screened out separately. The objective is to manufacture fine particles with a fine particle size distribution [51]. A graphic illustration of the overall development is described in Fig. 17.

The final particle sizes vary from 15 to 75 μm. While an existing material has a typical size of the particle around 20 μm. The fine particles are then coated through electrostatic spray gun and have thickness of greater than 500 μm. Distinct traditional coatings, usually finds no difficulties with scorching or sagging.

All the manufacturing methods including extrusion, particle size sieve shakers must be such that the softening point which is the most critical parameter should be quite high. Similarly, the other most significant character is the Tg (glass transition) of material and it must fall above 50 °C. The hard resin is then added to the milling stage and thus its significance is limited if meet all above parameters [52].

Film forming of powder coatings brings place by a melt process. The low viscosity after melting is mostly beneficial as the required viscosity for stoving application. This feature of best fit levelling parameters is saliant feature in powder coatings. This is the most excellent condition for optimal levelling [53].

5.6 Composition of Powder Coatings and Superior Characteristics

5.6.1 Thermoplastic Powder Coatings

One of the comparatively cheaper coatings are categorically known as TPC (Thermoplastic powder coatings). These are basically functional coatings and are not too much resistant to chemical and solvent attack as compared to those powder coatings which are chemically crosslinked. Most of the application area of thermoplastic powder coatings are domestic appliances [54].

Polyethylene (P.E) powders are considered most cost-effective finished coatings. P.E powders are primarily utilized by fluid bed sintering. Similar to other application, films are formed at the higher temperature usually around 270–410 °C. If there is special requirement of substrate, specific primer for adhesions is incorporated. There are different primers for different substrate to be applied for coatings, so specific primers for adhesion are applied before final coatings. One of the drawbacks for such coatings that these coatings are not weather resistant. There is another class of polymer-based powder coating originated from polyamide-12 and marketed as Polyamide powders. The method of producing and application is same as the normal powder coating techniques. These powders are also producing films of certain thickness at the temperature ranges from 285 to 415 °C. The resultant film produced from polyamide powder have weather resistant characteristics and also much more chemical resistant as compared to polyethylene powder coatings. Despite of the fact that coating film produced have certain drawback like absorption of moisture and yellowish in appearance, thermoplastic based powder coatings are still the most widely used material. However, they are the utmost frequently used thermoplastic powder coatings. Powder coatings based on PVC are produced from polyvinyl chloride and plasticizer as well. The films are rather elastic, nevertheless are not yellowing resilient. PVC powders are low-cost. They are useful by electrostatic spraying and produce films at 160–200 °C. For various substrates, it is essential to practice an adhesion primer [55].

Powder coatings comprising ethylene–vinyl alcohol copolymers are primarily applied by fluid bed sintering, and their film forming temperatures vary from 200 to 400 °C. The films are recognized by excellent flexibility and adhesion. They are somewhat resistant to corrosion and weathering. Polyester powder coatings comprise of high-molecular, linear, aromatic polyesters. They are generally utilized by electrostatic spraying and form films at 200–240 °C. The films exhibit excellent flexibility, high adhesion, and demonstrate sufficient resistance to water and weathering [56].

6 Functional Polymeric Coatings

In multilayer techniques, functional coatings are typically effective at the several interfaces including, polymer-air, polymer-substrate (wood, metal, glass, or plastic), polymer-liquid (oil or water), and eventually polymer-polymer. Therefore, coatings are made up of fringe structures that are constantly interacting with the environment and users. These structures are aggressively attacked by a variety of factors such as chemicals, bacteria, rain, dust, oils, and wear, which gradually reduces the performance of the ingredients and shortens their shelf life. In this context, efficient coatings have very much in common with the boundary structures of living organisms [57].

6.1 *Self-healing Polymeric Coatings*

Coatings are frequently employed around a substantial surface area and are hence the extremely possible setting for the initial harms to happen. Numerous self-healing methods have been described for polymers and/or polymeric coatings, which can be categorized in different ways, corresponding to the treatment process: variable versus irrevocable, and intrinsic versus extrinsic healing. The third category seems to be gathering substantial compromise in the coatings field. An extrinsic approach includes the addition of external factors to the technique or preparation, which are not a part of the major polymer group (e.g., microvascular networks or filled capsules). An intrinsic method utilizes curing agent's characteristic to the material which are usually a portion of the system. Mutually methodologies have benefits and drawbacks. The external healing model permits a limited reaction and delivers specific elasticity in opting different external factors, but it is fairly inadequate for numerous healing outcomes. The fundamental attempt may permit several healing and a renovation of the materials to a point which is incredibly near to the unique state [58].

6.2 *Antibacterial Polymeric Coatings*

Antibacterial and antimicrobial additives are agents efficient enough to eradicate the toxic microorganisms. Antibacterial agents could be classified into further two subcategories on the basis of extent of activity and amount of residue produced: The first category related to killing of bacteria with quick action leaving behind no residue and that's why termed as residue free antibacterial mechanism. The peroxide, aldehyde, alcohols and halogen releasing compounds are excellent example of first class of antibacterial compound. The next class comprises mostly of fresher complexes that leave long-acting residues on the surface to be sanitized and thus have a persistent action and are referred to as residue-producing. Some of the emerging examples are of this group are triclosan, bisphenols, benzalkonium chloride, mercury and silver compounds, tri-chlorocarbon, quaternary ammonium (QA) compounds [59].

6.3 Novel Marine Antifouling Coatings (AF)

The type of coating utilized in special application like marine application are prone to fouling. In order to eradicate the fouling problem, Innovative antifouling coatings are developed and broadly categorized into two sub classes, one is chemically active and termed as biodegradation and other is chemically inert and termed as bio-prevention antifouling techniques. In first class of antifouling coatings, usually special coatings are developed based on enzymes and this is most advance form of the coatings. The later strategy is proposed based on the viewpoint that prevention is preferable to cure. The mechanism of eradication of biofouling from marine is to kill the existing one attached on the surface and also prevent the bacteria to deposit of the surface as well. Chemically inert AF strategies are produced by tuning the surface tensions and/or surface charges, mimicking natural surface topographies, and mixtures of these approaches [60].

6.4 Poly (Zwitterionic) Non-fouling (NF) Coatings

The field of coatings is emerging day by day and in recent innovation, surface coatings based on zwitterionic have taken much attention for the utilization of these coatings as anticoagulant, anti-bacterial purposes. anticoagulant, biomedical diagnostic agent, as well as ultra-filtration membranes. With the recent innovative approaches these coatings are versatile in nature specially, formation of a hydration layer by zwitterionic polymers or mixed charged groups, as depicted in Fig. 18. Generally, there are two types of zwitterionic polymer. One has both positively and negatively charged groups on the same monomer. These include phosphorylcholines and polybetaines. The other has differently charged components on separate monomers, creating a mixed charge complex. These are known as polyampholytes [61]. Polybetaines can be further divided into three major groups based on the negatively charged groups: sulfonate-betaines (SB), carboxylate-betaines (CB), and phosphonate-betaines (PB) (Fig. 19).

These coatings are not only resistant to chemical but also towards cells adsorption, numerous proteins for example, fibrinogen, serum protein, platelet, lysozyme, bacteria (*Staphylococcus epidermidis*, *Pseudomonas aeruginosa*, *Escherichia coli*), and fibroblasts [62].

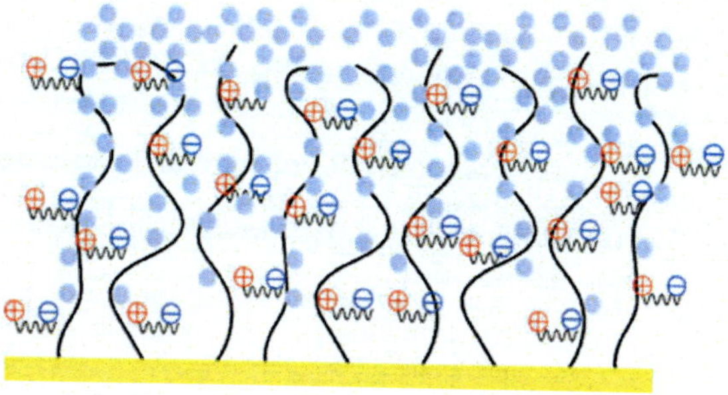

Fig. 18 Illustration of chain hydration and chain flexibility of zwitterionic polymers

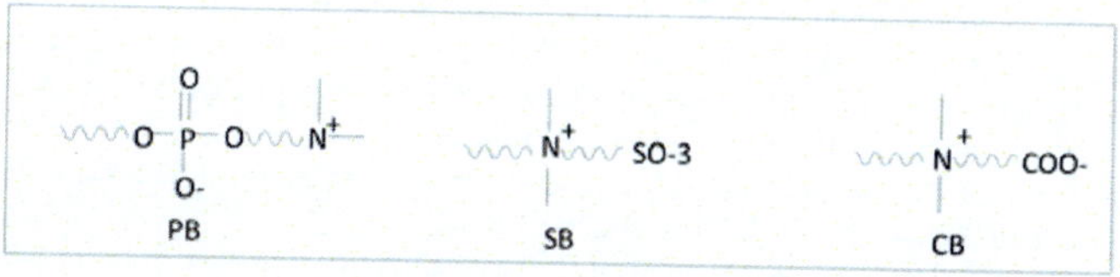

Fig. 19 Chemical structures of phosphonates-betaines (PB), sulfonate-betaines (SB), and carboxylate-betaines (CB)

References

1. Modern Coating Technology Wood Coatings A Sustainable Approach to Modern Coating Technology
2. Das, A., Mahanwar, P.: A brief discussion on advances in polyurethane applications. Adv. Ind. Eng. Polym. Res. **3**(3), 93–101 (2020). https://doi.org/10.1016/j.aiepr.2020.07.002
3. Diniz, F.B., De Andrade, G.F., Martins, C.R., De Azevedo, W.M.: A comparative study of epoxy and polyurethane based coatings containing polyaniline-DBSA pigments for corrosion protection on mild steel. Prog. Org. Coat. **76**(5), 912–916 (2013). https://doi.org/10.1016/j.porgcoat.2013.02.010
4. Ates, M.: A review on conducting polymer coatings for corrosion protection. J. Adhes. Sci. Technol. **30**(14), 1510–1536 (2016). https://doi.org/10.1080/01694243.2016.1150662
5. Samadzadeh, M., Boura, S.H., Peikari, M., Kasiriha, S.M., Ashrafi, A.: A review on self-healing coatings based on micro/nanocapsules. Prog. Org. Coat. **68**(3), 159–164 (2010). https://doi.org/10.1016/j.porgcoat.2010.01.006
6. Iscen, A., Forero-Martinez, N.C., Valsson, O., Kremer, K.: Acrylic paints: an atomistic view of polymer structure and effects of environmental pollutants. J. Phys. Chem. B **125**(38), 10854–10865 (2021). https://doi.org/10.1021/acs.jpcb.1c05188
7. Lestido-Cardama, A., et al.: Characterization of polyester coatings intended for food contact by different analytical techniques and migration testing by LC-MSn. Polymers (Basel) **14**(3) (2022). https://doi.org/10.3390/polym14030487
8. Müller, B., Poth, U.: Coatings Formulation, vol. 36, no. 5 (2007)
9. Xu, D., Liu, J., Ma, T., Zhao, X., Ma, H., Li, J.: Coupling of sponge fillers and two-zone clarifiers for granular sludge in an integrated oxidation ditch. Environ. Technol. Innov. **26**, 102264 (2022). https://doi.org/10.1016/j.eti.2021.102264

10. Chundawat, T.S., Verma, N., Vaya, D.: Development in synthesis and coating applications of polyurethane. J. Chil. Chem. Soc. **66**(2), 5142–5148 (2021). https://doi.org/10.4067/S0717-97072021000205142
11. Makhlouf, A.S.H.: Conventional and advanced coatings for industrial applications: an overview. Nanocoat. Ultra-Thin Film. Technol. Appl. 159–181 (2011). https://doi.org/10.1016/B978-1-84569-812-6.50006-3
12. Wu, L., Baghdachi, J.: Functional Polymer Coatings: Principles, Methods, and Applications (2015)
13. Olabintan, A.B., Ahmed, E., Al Abdulgader, H., Saleh, T.A.: Hydrophobic and oleophilic amine-functionalised graphene/polyethylene nanocomposite for oil–water separation. Environ. Technol. Innov. **27**, 102391 (2022). https://doi.org/10.1016/j.eti.2022.102391
14. Fernández-álvarez, M., Velasco, F., Bautista, A., Lobo, F.C.M., Fernandes, E.M., Reis, R.L.: Manufacturing and characterization of coatings from polyamide powders functionalized with nanosilica. Polymers (Basel) **12**(10), 1–20 (2020). https://doi.org/10.3390/polym12102298
15. Armelin, E., Oliver, R., Liesa, F., Iribarren, J.I., Estrany, F., Alemán, C.: Marine paint fomulations: conducting polymers as anticorrosive additives. Prog. Org. Coat. **59**(1), 46–52 (2007). https://doi.org/10.1016/j.porgcoat.2007.01.013
16. Felton, L.A.: Mechanisms of polymeric film formation. Int. J. Pharm. **457**(2), 423–427 (2013). https://doi.org/10.1016/j.ijpharm.2012.12.027
17. Yang, Z., Peng, H., Wang, W., Liu, T.: Crystallization behavior of poly(ε-caprolactone)/layered double hydroxide nanocomposites. J. Appl. Polym. Sci. **116**(5), 2658–2667 (2010). https://doi.org/10.1002/app
18. Mattar, H., et al.: Nitrocellulose: Structure, Synthesis, Characterization, and Applications, vol. 15, no. 3, pp. 1–15 (2020)
19. Talbert, R.: Paint Technology Handbook (2013)
20. Karger-Kocsis, J.: Paints, Coatings and Solvents, vol. 51, no. 4 (1994)
21. Paints, Pigments, and Industrial Coatings
22. Raju, K.V.S.N., Chattopadhyay, D.K.: Polyester Coatings for Corrosion Protection (2008)
23. Puhan, S.S., Central, A.E.: Chemistry project no. 6602984 (2014)
24. Sallis, J.G.M.: Polymer Coatings (1976)
25. Barroso, G., Li, Q., Bordia, R.K., Motz, G.: Polymeric and ceramic silicon-based coatings—a review. J. Mater. Chem. A **7**(5), 1936–1963 (2019). https://doi.org/10.1039/c8ta09054h
26. Zhang, W., Liao, L.P., Zhao, Y.: Incorporating microcapsules in smart coatings for corrosion protection of steel. In: Handbook of Smart Coatings for Materials Protection, pp. 287–306 (2014). https://doi.org/10.1533/9780857096883.2.287
27. Kausar, A.: Polyurethane nanocomposite coatings: state of the art and perspectives. Polym. Int. **67**(11), 1470–1477 (2018). https://doi.org/10.1002/pi.5616
28. Defeyt, C., Langenbacher, J., Rivenc, R.: Polyurethane coatings used in twentieth century outdoor painted sculptures. Part I: comparative study of various systems by means of ATR-FTIR spectroscopy. Herit. Sci. **5**(1), 1–12 (2017). https://doi.org/10.1186/s40494-017-0124-7
29. Li, G., Wei, Y., Ren, R.: Preparation of high-temperature-resistance silicone paints. Adv. Mater. Res. **602–604**, 1676–1679 (2013). https://doi.org/10.4028/www.scientific.net/AMR.602-604.1676
30. Amrutkar, S., More, A., Mestry, S., Mhaske, S.T.: Recent developments in the anti-graffiti coatings: an attentive review. J. Coat. Technol. Res. (2022). https://doi.org/10.1007/s11998-021-00580-z
31. Selim, M.S., et al.: Recent progress in marine foul-release polymeric nanocomposite coatings. Prog. Mater. Sci. **87**, 1–32 (2017). https://doi.org/10.1016/j.pmatsci.2017.02.001
32. Liu, Y.: Research on a superhydrophobic coating of highly transparent wear-resistant inorganic/organic silicon composite resin. (2021)
33. Fiume, M.M., et al.: Safety assessment of nitrocellulose and collodion as used in cosmetics. Int. J. Toxicol. **35**(Supplement 1), 50S-59S (2016). https://doi.org/10.1177/1091581816651607
34. Bubat, A., Scholz, W.: Silicone additives for paints and coatings. Chimia (Aarau) **56**(5), 203–209 (2002). https://doi.org/10.2533/000942902777680513

35. Zielecka, M., Rabajczyk, A., Cygańczuk, K., Pastuszka, Ł, Jurecki, L.: Silicone resin-based intumescent paints. Materials (Basel) **13**(21), 1–18 (2020). https://doi.org/10.3390/ma13214785
36. Ramakrishnan, T., et al.: Study of various epoxy-based surface coating techniques for anti-corrosion properties. Adv. Mater. Sci. Eng. **2022** (2022). https://doi.org/10.1155/2022/5285919
37. Xu, Y., Tan, L., Li, Q., Zheng, X., Liu, W.: Sublethal concentrations of heavy metals Cu2+ and Zn2+ can induce the emergence of bacterial multidrug resistance. Environ. Technol. Innov. 102379 (2022). https://doi.org/10.1016/j.eti.2022.102379
38. Lomelí-Rodríguez, M., Corpas-Martínez, J.R., Willis, S., Mulholland, R., Lopez-Sanchez, J.A.: Synthesis and characterization of renewable polyester coil coatings from biomass-derived isosorbide, FDCA, 1,5-pentanediol, succinic acid, and 1,3-propanediol. Polymers (Basel) **10**(6), 1–19 (2018). https://doi.org/10.3390/polym10060600
39. Gottsegen, M.D.: The Painter's Handbook, pp. 199–236. (2006)
40. Yanping, Y.: The development of polyurethane. Mater. Sci. Mater. Rev. **1**(1), 1–8 (2018). https://doi.org/10.18063/msmr.v1i1.507
41. Du, Z., Wen, S., Wang, J., Yin, C., Yu, D., Luo, J.: The Review of Powder Coatings, pp. 54–59 (2016)
42. Koolivand, A., et al.: The effect of petroleum hydrocarbons concentration on competition between oil-degrading bacteria and indigenous compost microorganisms in petroleum sludge bioremediation. Environ. Technol. Innov. **26**, 102319 (2022). https://doi.org/10.1016/j.eti.2022.102319
43. Adamu, A.A., Muhamad Sarih, N., Gan, S.N.: Thermal and anticorrosion properties of polyurethane coatings derived from recycled polyethylene terephthalate and palmolein-based polyols. R. Soc. Open Sci. **8**(4) (2021). https://doi.org/10.1098/rsos.201087
44. Tile-adhesives.pdf
45. Lee, Y.J., Lei, Z.: Wastewater treatment using microalgal-bacterial aggregate process at zero-aeration scenario: Most recent research focuses and perspectives. Bioresour. Technol. Reports, vol. 17, no. December 2021, p. 100943 (2022). https://doi.org/10.1016/j.biteb.2021.100943
46. Zafar, S.: Waterborne epoxy based coating materials. Am. Sci. Res. J. Eng. Technol. Sci. **50**(1), 133–154 (2018)
47. Grubbs, R.B.: What is a Polymer ?, vol. 9000, pp. 1–18 (2014)
48. Water-Based Paint Formulations, vol. 4. N. Publications
49. Lan, P., Nunez, E.E., Polycarpou, A.A.: Advanced polymeric coatings and their applications: Green tribology. Encycl. Renew. Sustain. Mater. 345–358, (2020). https://doi.org/10.1016/b978-0-12-803581-8.11466-3
50. dePolo, G., Walton, M., Keune, K., Shull, K.R.: After the paint has dried: a review of testing techniques for studying the mechanical properties of artists' paint. Herit. Sci. **9**(1), 1–24 (2021). https://doi.org/10.1186/s40494-021-00529-w
51. Emery, J., Stone, G.: Chapter 1: Introduction and literature review. Libr. Technol. Rep. **49**(2), 5–9 (2013)
52. Iñiguez-moreno, M., Ragazzo-sánchez, J.A., Calderón-santoyo, M.: An extensive review of natural polymers used as coatings for postharvest shelf-life extension: trends and challenges. Polymers (Basel) **13**(19) (2021). https://doi.org/10.3390/polym13193271
53. Pradhan, S., Pandey, P., Mohanty, S., Nayak, S.K.: Insight on the chemistry of epoxy and its curing for coating applications: a detailed investigation and future perspectives. Polym. Plast. Technol. Eng. **55**(8), 862–877 (2016). https://doi.org/10.1080/03602559.2015.1103269
54. Rangel, E.C., et al.: Cell adhesion to plasma-coated PVC. Sci. World J. **2014** (2014). https://doi.org/10.1155/2014/132308
55. Bauer, P., & Buettner, A.: Characterization of odorous and potentially harmful substances in artists' acrylic paint. Front. Public Heal. **6**, 1–11 (2018). https://doi.org/10.3389/fpubh.2018.00350
56. Kumarasinghe, K.G.U.R. et al.: One-pot reducing agent-free synthesis of silver nanoparticles/nitrocellulose composite surface coating with antimicrobial and antibiofilm activities. Biomed Res. Int. **2021** (2021). https://doi.org/10.1155/2021/6666642

57. Lambourne R., Strivens, T.A.: Paint and Surface Coatings Theory and Practice, p. 800 (1999)
58. Halladay, J.R.: Elastomeric coatings. KGK Kautschuk Gummi Kunststoffe **60**(5), 257–260 (2007). https://doi.org/10.1016/0042-207x(64)90583-4
59. Bandeira, R.M., van Drunen, J., TremiliosiFilho, G., dos Santos, J.R., de Matos, J.M.E.: Polyaniline/polyvinyl chloride blended coatings for the corrosion protection of carbon steel. Prog. Org. Coat. **106**, 50–59 (2017). https://doi.org/10.1016/j.porgcoat.2017.02.009
60. Krishnaveni, D., Kannan, P., Senthilkumar, A.N., Raja, K.: Safe disposal of phosphate for eutrophication control by Redgram stalk biochar with subsequent power generation. Environ. Technol. Innov. **27**, 102389 (2022). https://doi.org/10.1016/j.eti.2022.102389
61. Bergamasco, S., Tamantini, S., Zikeli, F., Vinciguerra, V., Mugnozza, G.S., Romagnoli, M.: Synthesis and characterizations of eco-friendly organosolv lignin-based polyurethane coating films for the coating industry. Polymers (Basel) **14**(3) (2022). https://doi.org/10.3390/polym14030416
62. Moon, J.I., Lee, Y.H., Kim, H.J., Noh, S.M., Nam, J.H.: Synthesis of elastomeric polyester and physical properties of polyester coating for automotive pre-primed system. Prog. Org. Coat. **75**(1–2), 65–71 (2012). https://doi.org/10.1016/j.porgcoat.2012.03.009

Smart Polymers and Composites

Iqra Abdul Rashid and Ayesha Afzal

Abstract This chapter covers the smart polymers and composites used for various applications like fuel cells, bio-medical applications, temperature and pH-responsive polymers, membranes, and nanocomposites. Smart polymers and composite popularity is increasing for various reasons like the polymers can be designed as environmentally friendly, stimuli-responsive like green polymers, smart polymeric coatings, biopolymers for cleaning dyes from water at the end of adsorbent, and energy efficient as a fuel cell and smart filtration process as a membrane. Other than conventional membrane, stimuli-responsive membrane materials are also discussed. The use of biocompatible polymer and biodegradable polymer materials for various applications is also included in the chapter.

1 Introduction

Nature is the best source of inspiration for the development of new materials. The properties of these materials also need to be tuned according to our application areas/requirements. Smart polymers undergo chemical or physical changes in response to environmental variations and are therefore also called stimuli-responsive polymers. Some of the most common stimuli for these polymers include pH, temperature, magnetic field, biological molecules, light intensity, etc. The polymers respond to these stimuli either by collapsing, swelling, altering their hydrophilicity, etc. In addition, polymers can be functionalized through pre-polymerization or post-polymerization by the addition of functional molecules in their chemical structure. Smart polymer and composite have also been used as energy-efficient materials in the form of batteries, and fuel cells. They are replacing typical batteries by introducing polymer composite membranes as electrolyte that is safe as compared to liquid electrolytes. The use of polymer composite membrane is not limited to fuel cells but it has been used for filtration processes like micro-scale filtration, nano-scale filtration as well. The membrane technology is useful for water filtration plants, dying industry.

I. A. Rashid · A. Afzal (✉)
Department of Materials, National Textile University, Faisalabad, Pakistan
e-mail: ayesha.afzal@ntu.edu.pk

K. Shaker and A. Hafeez (eds.), *Advanced Functional Polymers*, Engineering Materials,
https://doi.org/10.1007/978-981-99-0787-8_5

Smart polymer composites have been expanded to nanocomposite fabrication that has played a significant role in field of bio-technology, electronics, robotics, & actuators [1].

2 Temperature-Responsive Polymers for Hydrogels Applications

Temperature-responsive polymers can respond to spontaneous variations in temperature. This feature allows the use of temperature-responsive polymers in medical applications, e.g., the increased temperature of inflamed tissue as well as day and night cycles. Different techniques can be used for the preparation of these polymers. Among these techniques three major classes are:

1. Liquid crystalline polymers
2. Shape memory polymers
3. Responsive polymer solutions

Liquid crystalline polymers have a combination of a glassy state, liquid crystalline phase, and isotropic rubbery phase. Polymers having mesogens with anisotropic order are present in the liquid crystalline phase. Polymer networks having side chains can be prepared with a chiral nematic liquid crystalline phase, which is utilized in LED screens. These polymers can be used for the preparation of thermochromic materials.

Shape memory polymers are thermoplastic elastomers that consist of a hard phase having high (Tg; Tg,1). A second switching phase has intermediate glass transition temperature (Tg,2) and melting temperature which allows the temperature-responsive behavior. When these shape memory materials are heated above their Tg they deformed into a permanent shape. A temporary shape can be formed when the material is heated between two transition temperatures. By cooling below the switching temperature, the shapes can be frozen. But this can be transformed back into a permanent shape after heating above switching temperature. Although these materials are thermo-responsive after every switching cycle, for which they have to be reprogrammed. A shape memory polymer with one broad Tg can be programmed into four independent states. For example, shape memory polymers made from a network of liquid crystalline polymers show a wide colors range between orange and blue (Fig. 1).

The third thermo-responsive polymers are the one that undergoes a liquid–liquid phase transition. That phase transition seems like a change from a cloudy solution to a clear solution. This phase transition is called cloud point temperature (T_{CP}). Upon heating when phase separation occurs this transition is called lower critical solution temperature (LCST) and the reverse pattern is called upper critical solution temperature (UCST) (Fig. 2).

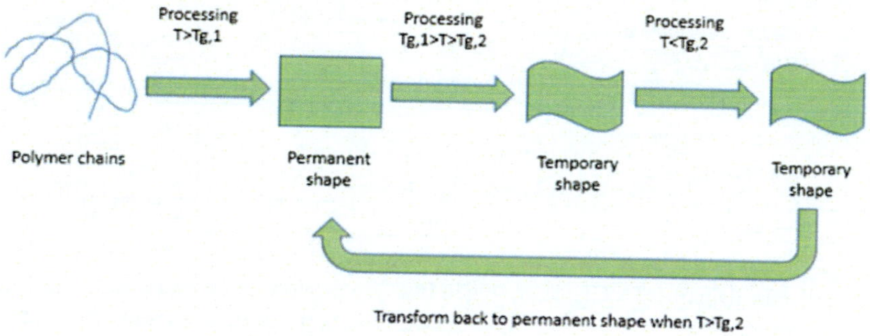

Fig. 1 Shape-memory polymer's thermo-responsive behavior

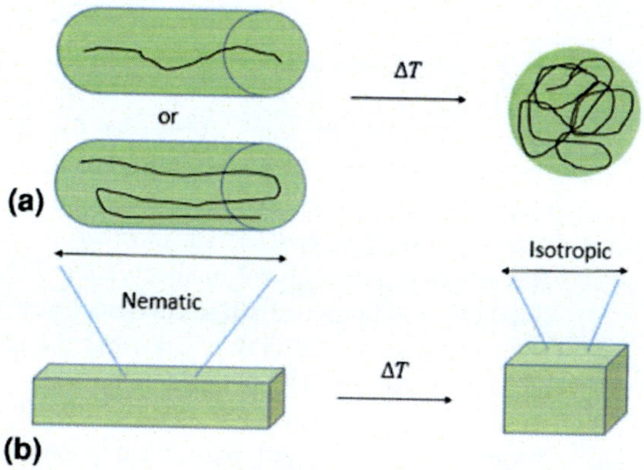

Fig. 2 **a** Heat application on the main chain of LC polymer to alter the conformation from extended phase to malformed phase **b** Macroscopic phase transition of nematic-isotropic

2.1 *Types of Temperature-Responsive Polymers*

Normally, the polymers dissolved in water show LCST behavior. It is because the total heat of hydration is less and for hydrating water molecules, the loss of entropy is high when the temperature is increased. However, several polymers show LCST behavior in water at some specific temperatures from 0–100 °C. That is the reason, a subtle balance of hydrophobic and hydrophilic functional groups is necessary for the polymer structure.

2.1.1 Polyacrylamides and Poly Vinyl Amides

Poly (N-isopropyl acrylamide) (PNIPAM) is a mostly studied thermoresponsive polymer in an aqueous solution. The use of PNIPAM in biomedical applications as LCST is not just because of the working temperature window of ~32 °C but also for its quick response. Other than PNIPAM, the LCST behavior of poly(N-vinyl

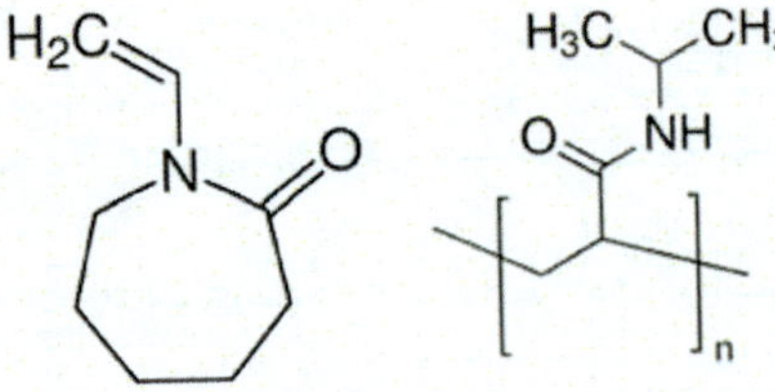

Fig. 3 Structures of poly(N-vinyl caprolactam) and poly(N-isopropyl acrylamide) (PNIPAM)

caprolactam) (PVCL) as shown in Fig. 3, was seen at approximately ~31 °C. Furthermore, PVCL is also biocompatible which makes it a suitable candidate for biomedical applications. However, PNIPAM is set as the gold standard among thermoresponsive polymers, for different biomedical applications.

2.1.2 Poly Oligo Ethylene Glycol Methacrylates

Other than PNIPAM, some alternatives for LCST polymers having better properties are discussed below. The most important properties are (1) lowering the Tg to minimize or remove glassy phase formation and reduce the formation of hysteresis and (2) easy synthesis through CRP. The most widely used grade alternative to PNIPAM is the poly oligo ethylene glycol methacrylates (POEGMA) having a poly(meth)acrylate (PMMA) backbone with oligo ethylene glycol as side chains. The thermoresponsive behavior of such POEGMAs describes the efficiency of the TCP by changing the number of ethylene glycol repeated units.

Similar to poly acrylamides and polyvinyl amides, POEGMA can be synthesized through free radical polymerization, anionic polymerization, and CRP, while the latter two methods show the proper end-groups in polymer structures.

2.2 Applications of Thermoresponsive Polymers

A dual sensor has been developed by combining an LCST polymer with pH-responsive solvatochromic dye. This sensor allows the determination of pH and temperature with UV measurement. The LCST transitions change the polymer confirmation, and this change is used to control the side change functionalities. The strong PNIPAM nanoparticles binding was achieved below the T_{CP} as compared to their warped state, which shows the availability of the mannose units.

Thermoresponsive polymers are also used for the preparation of hydrogels. Extremely concentrated thermoresponsive polymer solutions show gelation. Further heating enhances the network structure and dehydration due to heat. After dehydration, the chains solidify as a gel. In addition, a thermoresponsive poly(isocyanide) bearing ethylene glycol functionalized peptidic side chains were studied to undergo LCST temperature-responsive hydrogenation at a concentration below 0.006 wt%.

The effective hydro-gelation mainly depends on the stiffness of polymer chains. The helical polymer chains bundle with each other via physical crosslinking and dehydrating side chains [1].

3 Conductive Polymers for Bioelectronic Applications

Electrically conductive polymers found great interest in bioelectronic applications. They have an excellent ability to use in numerous devices which transform electronic signals into ionic signals and ionic to electronic signals. Their soft nature, composition, and charge transport make them suitable for living tissue and solid substrates.

Different conductive polymers can be used for this purpose but poly (3,4-ethylene dioxythiophene): poly(styrene sulfonate) (PEDOT:PSS) is most widely used in industries [2, 3]. Because of its excellent water dispersibility, better stretchability, and high stability we can easily prepare films from this material. This is used in many applications of organic electronics as a hole conducting layer, transparent conductive oxides (TCO), and electrochromic layer in different devices from organic photovoltaic devices (OPVs), organic light-emitting diodes (OLEDs), and electrochromic. In the past (PEDOT: PSS) is used in electrical equipment like electrodes for electrophysiology, biosensors, electronic textiles, electronic skin, etc. But this material faces some difficulties due to its limited electrical conductivity, flexibility, low biocompatibility, and functionality.

3.1 PEDOT: PSS Dispersion

These dispersions are made through oxidative polymerization of EDOT by using a solubilized PSS. By using this technique these prepared PEDOT insoluble particles are stabilized and doped with PSS. Also the complex particles of PSS and PEDOT are formed in which the chains of PEDOT are coupled with shells of PSS, which gives complete balance to the chains of PEDOT to distribute in suitable solvents. The mixture of PEDOT: PSS can be used for water diffusion because it is easily dissolved in organic polar solvents e.g. alcohols. This aqueous diffusion of PEDOT: PSS is used as conducting polymer for different industrial fabrication and applications. The aqueous dispersion of PEDOT: PSS is also used for electronic applications due to its high transparency, mechanical flexibility, and thermal stability. But when this solution is cast into a film its conductivity becomes low. To increase the conductivity secondary dopants are used nowadays which include salts, acids, polyols, and organic solvents like HCl, H_2SO_4, and DBSA [4–9]. Through this second dopant PEDOT: PSS has a range of electrical conductivities from 10^{-2} to 4600 S cm^{-1}.

When the film is formed then it is required to avoid delamination and dispersion of chains when in contact with water in an aqueous medium like biological fluids. To maintain the fastability water-soluble chemicals are added before the film formation. These chemicals are polyethylene oxide (PEO), polyvinyl alcohol (PVA), etc. The prepared films are further used for bioelectronic applications.

3.2 Applications of Biopolymer Diffusions

PEDOT: PSS is used in bioelectronics because of its electronic conductivity, organic nature, and reaction to electrical impulse (wettability, change in color, and volume). This is very good for electrochemical transistor (OECT) devices. OECT is better as compared to silicon field-effect transistors (FET)s because it gives a high current. This is because of the complete film contribution as charge transport in the channel, the same as FETs having thin layers together with an aqueous system. In addition, the OECT instruments are prepared as transparent sensors which are well-suited with an inverse microscope that is used in cell culture labs.

4 Polymer-Based Nano Sorbents for Dye Removal Applications

For the purification of water, the adsorption process is most commonly used due to its high reuse capability, simplicity, less sludge production, and simple operation. The selection of polymeric materials used as adsorbents for dye removal depends on the adsorption capacity, reusability, large volume of pores, adsorption capacity, and cost of the adsorbent. Biopolymer-based adsorbents like polysaccharides are more environmentally friendly than other adsorbents.

4.1 Types of Polymeric Adsorbents

4.1.1 Nano-magnetic Polymers (NMPs)

Nano-magnetic polymers (NMPs) have different applications like biosensors' practicality, magnetic storage capability, jet printing as magnetic inks, and use in dye molecule uptake having good adsorption capacity, high selectivity of certain molecules, and fast rate.

The core–shell formation in NMPs can be done through the Fe_3O_4 coated procedure with the addition of functional groups. The functional group incorporation increases its adsorbent rate, selectivity, and stability. The polymers are engrafted with magnetic nanoparticles which enhance the adsorption rate and thus help to extract toxic dye elements including cationic or anionic ones.

4.1.2 Polysaccharides

Polysaccharides are the most widely used adsorbents due to their certain properties like biodegradable, inexpensive, biocompatible, and renewable which makes them environmentally friendly materials. The superior properties of polysaccharides such as specific structure, chemical and physical properties, stability, and excellent reactivity are due to the presence of chemical groups like hydroxyls, acetamido, and amino functions in polysaccharides chains which makes them excellent adsorbents for many dyes.

4.1.3 Carbon-Based Adsorbents

Carbon-based materials have been widely used for synthetic dye removal from aqueous solutions.

The excellent properties of carbon-based materials include stability in thermal and chemical factors, easy recycling, high surface area, and pore volume. These adsorbents can be divided into four groups in terms of removal capacity and specific surface area, excellent (>1000 mg/g), good (500–1000 mg/g), average (100–500 mg/g), and weak capacity (<100 mg/g). The dye adsorption efficiency is affected due to some parameters such as the size of particles and acidity of the surface through which the functionality will be completely changed, which changes the carbon molecules and dye behavior.

4.1.4 Bio-based Adsorbents

There are different economic sorbents such as agricultural run-offs, natural materials, and microorganisms. But they often suffer from less potential in adsorption processes. Therefore, their functionality can be improved by increasing the surface area and modifications in the functional group.

These bio-based adsorbents gain attention due to their biocompatibility, biodegradability, inexpensiveness, and stability in organic solvents. For the removal of cationic contaminants, biosorption was employed but for the removal of anionic contaminants, functionalization must be done [10].

5 Smart Polymer for Fuel Cell Applications

A fuel cell resembles a battery in many aspects; however, it is more useful than batteries when electrical energy is required for a longer period. It is because of the continuous supply of oxygen from external sources whereas batteries contain a limited fuel supply. Fuel cells are potential energy conversion devices, and they work in hydrogen or hydrogen-rich fuels. They produce low CO_2 emissions. The popular type is proton exchange membrane fuel cells (PEMFC). This type of fuel cell exhibits extraordinary attributes i.e. high energy conversion efficiency, limited fuel crossover effects, and high open circuit potential. Moreover, they efficiently undergo electro oxidation of organic molecules at low temperatures these molecules are ethanol, formic acid, and methanol. Alternatively, microbial fuel cells (MFCs) directly transform organic waste into electricity via reactions like microbial/enzymatic/abiotic and microbially catalyzed anodic. The advancements in fuel cell technology are due to the use of polymer composite. This type of cell is fabricated around an ion-conducting membrane. The electrodes consist of catalyzed carbon. This type of fuel cell is called a solid polymer electrolyte fuel cell (PEFC). It functions very well but the cost of the system is higher than the usual batteries. The physicochemical properties of the electrolyte membrane were improved by a polymer-based composite system.

5.1 *Working Mechanism of Fuel Cell*

The fuel cell is a device to generate electricity through a chemical reaction. It consists of three components; cathode, anode, and electrolyte. The cathode and anode are collectively called electrodes. The electrodes are responsible for the production of electricity. The electrolyte carries electrically charged particles between the electrodes and needs a catalyst to speed up the reaction process.

5.2 *Types of Fuel Cells*

- Polymer Electrolyte Membrane Fuel Cells (High temperature/low temperature).
- Direct-Methanol Fuel Cells.
- Alkaline Fuel Cells.
- Phosphoric Acid Fuel Cells.
- Molten Carbonate Fuel Cells.
- Solid Oxide Fuel Cells.
- Combined Heat and Power Fuel Cells.
- Regenerative or Reversible Fuel Cells.

5.3 *Polymeric Materials as Electrolytes*

Numerous polymeric materials have been used as solid electrolyte membranes. It includes well-known material; a perfluorinated sulfonic acid membrane known as Nafion®. The other materials are polystyrene sulfonic acid, commercially sulfonated poly (arylene ether ketone), [poly(2,2-m-(phenylene)-5,5-bibenzimidazole] (PBI), poly-3-methyl thiophene (PMT) sulfonated poly (phenylene oxide)s, poly(phenylquinoxaline)s, and poly(phosphazene), poly(3,4-ethylene dioxythiophene) (PEDOT) polytetrafluoroethylene-reinforced perfluorinated sulfonic acid. The composite membranes also include inorganic reinforcements like titania, silica, zirconia, alumina, zirconium phosphate, sulfonated silica, and sulfonated titania. The alkaline fuel cells contain Bronsted base functionalities such as oxygen, nitrogen, or sulfur in polyvinyl alcohol (PVA), polyethylene oxide (PEO), and chitosan.

The fuel cell membrane conductivity should be 10^{-2} S/cm or higher to gain coherent in situ area-specific resistances of 0.1–0.3 Ω cm^2 with a membrane thicknesses of 40–150 mm.

The sulfonated membrane operating temperature is approximately 80 °C, and the phosphoric acid doped operating temperature is 180 °C [11].

5.4 *Materials as Electrodes*

In advanced fuel cells, the most popular category is Conducting Polymer Nanomaterials Hybrid (CPNH) electrocatalysts. It represents an exceptional kind of hybrid electro-catalysts. They provide high surface area along with the synergistic effect with metal nanoparticles/oxide to improve electrocatalytic activity and stabilize the cell. These materials include graphite, graphite felt, carbon paper, and carbon cloth as a Pt-free catalyst which is deposited on conducting polymers (CPs), etc. [12].

5.5 *Synthesis of Electrodes*

i. Chemical Polymerization
   (a) Photoreduction
   (b) Radiolysis
   (c) Wet Chemical Method
   (d) Chemical reducing agent
ii. Interfacial Polymerization
iii. In-situ Interfacial polymerization
iv. Electrochemical method
v. Electrochemical method via Galvanostatic electrode deposition

5.6 Synthesis of Membrane Electrolyte

The membrane preparation involves various reactions to obtain the specific structure, porosity, operating temperature, and thickness. The polymerization reaction can be typical in-situ polymerization and grafting polymerization, which helps build an active site in the polymer chain to obtain specific properties. The operating temperature higher than 100 °C uses acid–base polymers rather than using conventional Nafion. The Nafion loses proton conductivity in the absence of water while acid–base polymer membranes are water-independent. These membranes are made of functional sites like ether, imide, amide, imine, or alcohol groups. These groups create active sites to allow required proton conductivity. However, to achieve high proton conductivity strong acids are required. Regrettably, acid decreases the mechanical strength of polymers. Normally the temperature range is 150–200 °C for H3PO4-doped PBI-membrane because it offers high thermal, chemical, and mechanical stability at temperatures as high as 200 °C. PBI is an amorphous thermoplastic polymer with a glass transition temperature above 420 °C. Doping PBI with H_3PO_4 leads to the absorption of two H_3PO_4 molecules per monomeric unit of the polymer, resulting in the protonation of the imine group at the imidazole ring forming the corresponding salt [13].

6 Polymer-Based Smart Membrane Composite Applications

A polymer-based membrane can be defined as a thin semipermeable barrier. The membranes serve different applications i.e., water filtration, and dye removal, and act as a barrier between two gaseous phases and electrolyte. The electrolytic membranes are discussed in the previous section. The major application of membrane is for phase separation medium. We can say that polymeric membranes are a selective barrier system, and they can be used for water purification and drug purification to remove pollutants from wastewater streams, or for the reverse-osmosis process. The polymer-based membranes are more popular than inorganic membranes because of the versatility of their applications. They are preferable systems on an industrial scale owing to their pore-size formations, selectivity in phase separation, and ease of fabrication methods. Moreover, they possess flexible operating parameters at a low cost than inorganic membranes, ease of upscaling or downscaling, and a small chemical footprint. However, conventional polymeric membranes have a limitation, and there is a compromised relationship between selectivity and permeability. This problem has been addressed by using polymer-based nanocomposite membranes. Nanotechnology includes particles, nanotubes, nanofibers, and nanosheets. The membrane composite consists of a polymer matrix and nano reinforcements. These composite membranes exhibit high permeability & selectivity, modifying fouling behavior, eco-friendly, and energy-efficient.

6.1 Polymeric Materials for Membrane Preparation

The selection of polymers depends on the properties required for the functioning of the membrane. These properties are chain rigidity, stereoregularity, chain interaction, and polarity of functional groups. In terms of functional groups; a membrane should have a low binding affinity, and compatibility, and must follow the budget requirement.

The materials include cellulose acetate/triacetate, polybenzimidazole, polypiperazine for reverse osmosis, polyacrylonitrile (PAN), polyethersulfone (PES), polysulfone, polyvinylidene fluoride (PVF) for ultrafiltration, polyamide, polyphenol, polyol for nanofiltration, PTFE and PVF for distillation, and PVF, PTFE, PP, PE, PES and PEEK for micro filtration [14, 15].

There is a limited number of polymers that can be used to fabricate membranes and a maximum of these are hydrophobic. Which undesirably affects the purposeful properties of the membrane. These flaws can be minimized by the modification of membranes. The modification can be done in two ways; physical and chemical modification. It helps to alter the properties of the membrane. The altered properties play a vital role in the functioning of the membrane; selectivity and specificity.

6.2 Polymeric Membrane Preparation

Smart membranes are fabricated by surface grafting of stimuli-responsive polymers to obtain the required properties. The extent of grafting determines the characteristic properties. These types of membranes exhibit reversible changes in their physicochemical properties in relative environmental conditions; pH, ionic strength, temperature, and electromagnetic fields.

The second type of smart membrane is fabricated by surface modification. The smart polymers are grafted on the porous surface of the membrane. It creates nano-valves. It creates nano-valves on the membrane surface to enable the membrane to be stimuli-responsive [14–16].

6.3 Applications

- Water filtration
- Electrolytic membranes for fuel cell
- Tissue Engineering
- Biomedical Applications

7 Polymer-Based Functional Bio-nanocomposite for Smart Applications

Nanocomposite has ripened into the enormously active, dynamic, and intensifying field of research for emerging areas of energy production, paints and coating, structural materials, nutrition, cosmetics, personal care, agriculture, catalysts, lubricants, molecular computing, security printing, drug delivery, medical therapeutics, pharmaceuticals, and diagnostics. Usually, synthetic polymers are bio-incompatible and non-degradable. However, recent developments proved this statement wrong and we are at a pace where we can say bio-polymer does exist. Bio-polymers consist of monomers that are covalently bonded and form chain-like macromolecules. The degradation mechanism of biopolymers involves the chain scission and then complete molecule degradation through microorganisms into carbon dioxide and water. Bio-polymers are replacing single-use plastic to keep the environment clean and plastic-free.

7.1 Nano Materials

Nano-composite includes particles at the nanoscale to enhance or impart unique mechanical, physical, thermal, and electrical properties. These nanocomposites are carefully engineered with complex architecture to provide advanced applications. Moreover, polymer-based nanocomposites have gained noteworthy consideration because of their flexibility, specific stiffness, and low weight-to-strength ratio. More important are biocompatibility and biodegradability.

Nano-composites or nanofillers are of two types organic and inorganic materials. The organic nanofillers comprise natural clay, fibers, and polymer nanofibers. The inorganic nanofillers consist of carbon (nanofibers, graphene, fullerenes, and carbon nanotubes), metal oxides (TiO, ZnO, CuO, and FeO), and metals (Ag, Fe, and Au).

Recently, the advent of technology has led to the development of synthetic biodegradable polymers. It includes polyvinyl acetate (PVA), polycaprolactone (PCL), polyglycolic acid (PGA), polybutylene succinate (PBS), hydroxy-propyl-methyl-cellulose (HPMC), etc. These polymers show unique properties such as high gloss, clarity, tensile strength, durability, and flexibility.

Food packaging applications of plastics have a single-time utilization after that it remains on earth for several years. To resolve these problems bio-degradable food packaging has been introduced. It is made from chitosan, starch, cellulose, agar (carbohydrate source), whey protein, alginate, collagen, gluten, and gelatin (protein source).

7.2 *Categories of Bio-degradable Materials*

Bio-polymers are divided into four groups depending on the origin of the polymer [17].

1. Natural biopolymer extracted from biomass
2. Synthetic bio-polymers chemically synthesized from biomass
3. Synthetic biopolymers from microbial production or fermentation
4. Synthetic biopolymers are chemically synthesized from petroleum products.

Nowadays, food packaging also uses starch and its derivatives which are termed bio-nano composite. The packaging made from starch solely exhibits lower mechanical properties. That is the reason, starch must be used with reinforcement in form of a nanofiller or plasticizer [18]. The nanofiller can be in the form of nanorods like zinc oxide (ZnO) and nanoparticles like silver. The other nanofillers are montmorillonite (MMT), titanium oxide (TiO_2), etc.

These nanofillers exhibit distinguish properties and add a significant attribute to biodegradable material. For example, the antimicrobial property can be achieved by the addition of benzoic acid, propionic acid, and sorbic acid. Sensor property can be achieved by the addition of CNTs and MWCNTs. Bio-sensor responds to temperature changes, humidity, and oxygen level changes. It helps the consumer to track the quality of food and when the degradation process is going to start. It also detects microorganisms, toxic proteins, and spoilage of food or beverage. Some of the nanofillers act as oxygen scavengers to reduce or maintain the oxygen level for example TiO_2. There are some nanofillers that provide UV protection and most commonly ZnO serves the purpose.

7.3 *Classification of Polymer Nanocomposites*

7.3.1 Dimensional Classification of Nanofiller

Figure 4 shows Zero-dimensional (0D) nanofillers. These types of nanofillers are those particles that do not have any dimension and are in the range of 100 nm. They can be amorphous and crystalline in structure, and they can be in the form of polymeric materials, ceramics, and metallic. For example, quantum dots, gold, and silver with diameters ranging from 1–50 nm. One-dimensional (1D) nanofillers project dimension end to end in one direction. It ranges beyond nanometers. However, they are called nanomaterials because of their diameter which ranges from nanometers but has long lengths. The best examples are nano-tubes, nano-rods, nano-wires of metals, and metal oxides.

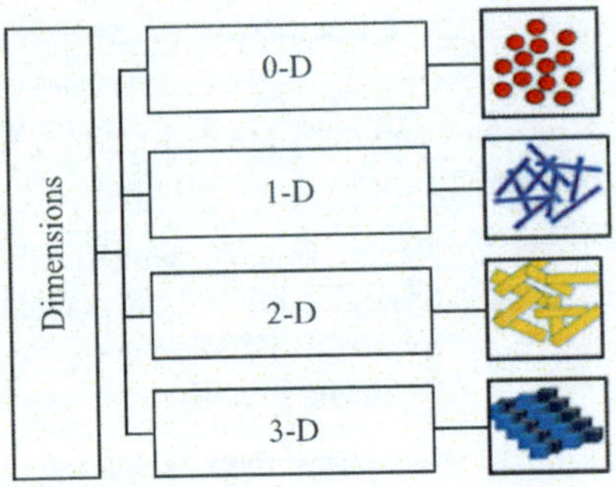

Fig. 4 Classification of nanoparticles based on dimensions

Two-dimensional (2D) nanofillers are those materials that exhibit their structure in two dimensions and are beyond nanometers. These materials are big and tremendously thin. It includes nanofilms, nanowalls, nanosheets, nanowires, nanowhiskers, and nanofibers; carbon nano-tubes and montmorillonite are a few remarkable examples.

Three-dimensional (3D) nanofillers exhibit dimensions in three axes. Their structure range beyond the nanometers however their unit structure range on the nanoscale. These materials comprise nano-clays, nano-granules, and equiaxed NPS; zeolites are a prominent example.

7.3.2 Types of Nanofillers

Figure 5 shows the category of nano-fillers. The nanoparticles can be spread easily and homogeneously onto the polymer matrix. The nano reinforcements can impart diverse properties in composites like flame retardancy and thermal stability. These nanoparticles include metal or metal oxide, silica, magnesium, iron oxide, titanium oxide, zinc, aluminum oxide, and zirconium oxide are characteristic examples. Another major example is metal-sulfide/polymer nanocomposites. They have earned significant attention because of their thermomechanical, electromechanical, and optical properties.

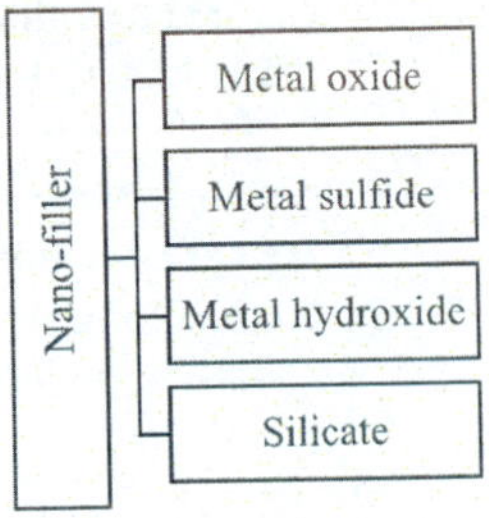

Fig. 5 Types of nanofillers

Nano-composites comprising metal sulfide e.g., CdS, ZnS, and HgS exhibit enhanced thermal stability. For flame retardant applications and thermal stability hydroxides of metal like aluminum hydroxide and magnesium hydroxide nanoparticles (NPS), and double hydroxides of zinc and alumina are truly remarkable. To enhance the mechanical properties silicate/polymer nanocomposites are on the priority list, this composite exhibit both mechanical and thermal properties. There are two methods to mix silica in the polymer matrix. The direct mixing method and solution/melt blending methods.

7.3.3 Type of Polymer Matrix

There are three types of polymer matrixes, thermoplastic, thermoset, and elastomers. The thermoplastic matrix can be softened and then molded under pressure and temperature and then hardened by cooling. They have tremendous properties such as remolding, recycling, moldability, high strength, and flexibility. Polystyrene (PS), polyethylene (PE), and polyvinyl chloride (PVC) are the best examples. The second type is thermosetting polymers. They are liquid at room temperature and upon heat, they start curing and hardening. After curing they exist in three-dimensional covalent-bonded structures. The cured thermoset matrix cannot be remolded. Examples are vulcanized rubber, epoxy resin, polyesters, and polyurethanes are examples of thermosetting polymers. The third category is elastomers, which are flexible, stretchable materials and are viscoelastic. Polybutadiene, silicone, epichlorohydrin, chloroprene, and polyacrylic rubber are examples of elastomer matrices.

7.4 Polymer Nanocomposite Preparation Methods

Polymer nanocomposites are prepared by different techniques:

(a) Insitu Polymerization
(b) Melt Processing
(c) Solution Blending/Casting method

7.4.1 In-Situ Polymerization

The in-situ polymerization technique is popular among other techniques and has been used for decades. It includes multiple ingredients; suitable solvents, and monomers followed by polymerization to yield polymer nanocomposite. This method allows the preparation of precise and controlled structures with diverse properties from their monomers. In this technique, homogenous dispersion plays a pivotal role and controls the size, shape, and morphology of nano-materials in a polymer matrix.

7.4.2 Melt Processing

The melt processing technique is suitable for thermoplastic materials. This technique is also known as green or environmentally friendly as it does not involve solvent. This is a striking technique because of its robustness. It is also an easy technique followed by extrusion and injection molding.

7.4.3 Solution Casting

The solution casting method is known for decades. This method is used to cast films with varying thicknesses. The polymer solution is prepared by dissolving it into a suitable solvent, and then the nanofillers are equally dispersed in the matrix to prepare a homogeneous mixture. Then the solution is cast over the mold as per the required shape to evaporate the solvent. This technique involves both organic and inorganic solvents.

7.5 Applications

- Active food packaging
- Edible films
- Medical application
- Drug delivery

8 pH-Responsive Polymers for Biomedical Applications

The pH-responsive polymers are the center of attention for the past decades. These smart polymers provide design flexibility for functional material fabrication. These polymers belong to polyelectrolytes, known for the structural changes in response to environmental pH. These polymers have either weak acidic or a basic group in their structures which is responsible for releasing or accepting protons with pH variation. These polymers find applications in the biomedical field, water remediation, personal care, and industrial processes. The biomedical field includes drug delivery, biosensors, and gene carriers. These polymers have wide scope to be used as anticancer drugs by minimizing the size to nanosize and maximizing the activity.

8.1 *Working Mechanism of pH-Responsive Polymers*

The working principle of pH-responsive polymers is based on the nature of the functional groups. They can be acidic or basic functional groups in polymer chain structures. These groups include carboxylic acids (–COOH), tertiary amines, sulfonic acids (–SO_3H), etc. These functional groups can ionize when pH changes. However, the electrostatic effect is wielded by head-to-head ionized groups which make the ionization process difficult. Therefore, the dissociation constant (Ka) value varies in monoacid or monobasic. The physical properties of such polymers can be designed according to the scope of application by manipulating charges along the backbone chain of the polymer. It results in electrostatic repulsive forces, and these forces increase the hydrodynamic volume of the polymer. The physical features that can be altered are the geometry of the chain, arrangement, dispersity, and volume of pH-responsive polymers. The change in hydrodynamic volume allows a transition from a tightly coiled state to an expanded state that can be explained by osmotic pressure. The pressure exerted by counter ions neutralized the charges. These changes are affected by the environment that amends electrostatic repulsion, such as pH, ionic strength, and type of counter-ions. The swelling and reduction in the volume of a polymer are linked with the acid and basic nature of the polymer respectively.

8.2 *Functional Groups in PH-Responsive Polymers*

The pH-responsive polymers are into groups:

- Acidic
- Basic

These polymers can be obtained from natural or synthetic sources. The naturally occurring pH-responsive polymers and multi-responsive polymers got serious attention in the past.

8.2.1 pH-Responsive Acidic polymers

These polymers consist of acidic functional groups in their structures. The acid groups are negatively charged therefore external pH variation brings changes in polymer chains having all negatively charged groups. This change allows hydrophilicity regulation in aqueous media.

These polymers are further classified on basis of acidic functional groups. For example; Carboxylic acid (poly (acrylic acid) and poly (*N*,*N*-dimethylamino ethyl methacrylate) (PDMAEMA)), Sulfonic acid (poly(2-acrylamide-2-methylpropane sulfonic acid) (PAMPS) and poly(4-styrene sulfonic acid), Sulfonamide groups.

8.2.2 pH-Responsive Basic Polymers

These polymers have basic functional groups in their structures. The variation in the quantity of positively charged groups occurs due to the peripheral change in pH value. Generally, ionization and de-ionization transition undergo at pH 7-11. The polycations used to prepare pH-responsive polymers are vinyl, acrylamide, pyridine derivatives, imidazole derivatives, methaacryl amide, methacrylates, and acrylates. In these cations, there are lone pair electrons accessible for proton bonding.

8.2.3 pH-Responsive Natural Polymers

Different biodegradable polymers fall into this category. The functional biopolymers have a significant property that they respond in all manner to peripheral stimulants. The slight variation occurs in response to erratic conditions. This transitional period stays until a critical point is reached. The most common polymers used for drug delivery systems are dextran, chitosan, alginate, and hyaluronic acid. The structures are shown in Fig. 6.

8.3 Synthesis of pH-Responsive Polymers

i. Emulsion Polymerization
   a. Miniemulsion Polymerization
   b. Microemulsion Polymerization
ii. Group transfer Polymerization
iii. Reversible Addition-Fragmentation Chain Transfer Polymerization (RAFT)
iv. Atom Transfer Radical Polymerization

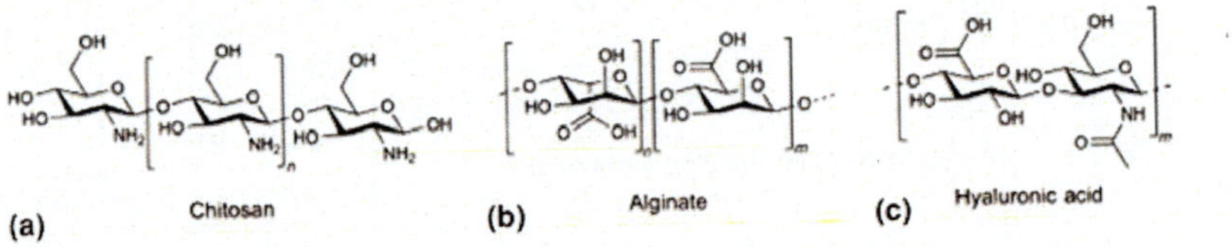

Fig. 6 a–c Chitosan, Alginate, and Hyaluronic acid, respectively are pH-responsive biopolymer structures that are used in drug delivery systems

8.4 Biomedical Application

The pH-responsive polymers have great potential areas. They can change the structure, volume, properties, chain configuration, etc. in response to changes in pH. These polymers are widely used for biotechnology (as drug delivery, gene carriers, and biosensors), nanotechnology, chromatography, coatings, and membranes. This chapter covers biomedical applications [16, 19].

8.5 Encrypted Polymers

There is a variety of toxins and viruses that have evolved multifunctional biomolecules, and this is a challenge that has to be addressed. To resolve this problem, researchers directed the cellular uptake and enhanced biomolecular transport to cytoplasm from the low pH endosomal compartment. The pH-responsive polymers that are bio-inspired and mimic the design of biology-based things are called “Encrypted Polymers”. It enhances cytosolic delivery by altering the structure of endosomal membranes. This technology is used in vaccines and for the delivery of bio-therapeutics [20].

9 Smart Polymeric Coatings for Biomedical Applications

There are numerous areas like biomedical and textile coatings, where smart or biopolymer coatings play an important role to impart anti-fogging, antibacterial, and self-healing properties.

Polymer coatings impart functionalities to the host materials. These coatings are applied to different materials like ceramics, nanoparticles, metals, and polymers. In the biomedical field, polymer coatings can be used for surface functionalization, corrosion resistance, and wear resistance. Smart polymers respond to different stimuli, such as light, temperature, magnetic field, and pH. Smart polymer-based nanocarriers are most widely used due to their controlled drug delivery, bioactivity, and sustained release. Polymer nanocomposites can be used in various applications by the incorporation of nanofillers. Numerous biopolymers like poly(2-methyl-2-oxazoline) (PMOXA), super-dispersed polytetrafluoroethylene (SPTFE), Polyvinylidene Fluoride (PVDF), Polymethyl Methacrylate (PMMA), Polypropylene (PP), and biodegradable Polycaprolactone (PCL) [21, 22].

9.1 *Methods of Biopolymer Coating*

Biopolymer coatings can be applied through different coating methods such as polymer brushes, Langmuir–Blodgett, layer-by-layer, and dip coating.

9.1.1 Layer by Layer (LBL)

In the layer-by-layer method, different types of charged polyelectrolytes are used to produce LBL coatings and film. The main purpose to use this process is its flexibility. Through this process different single-layer and multilayer coatings are produced for drug delivery and tissue engineering applications.

9.1.2 Langmuir–Blodgett (LB) Method

LB films are sometimes used as an alternative for the LBL process in which from the polymer solutions the substrates are drawing at a uniform speed. While in the LBL process, bulk solution molecules are coated on different surfaces, while in LB, molecules from the solution surface are coated on different substrates.

9.1.3 Polymer Brush Method

Polymer brush is an interesting and popular technique in which on the surface of the substrate, soft material is covalently tethered and has the potential to use in the biomedical field.

9.1.4 Plasma-Based Based Polymer Coatings

Plasma-based polymer coatings make it possible to have proper polymer coatings on surfaces for multiple applications. This method gives proper adhesion to substrates, such as ceramics and metals. In addition, this plasma coating gives possibilities to coat complex shapes.

9.2 *Biopolymer Coatings on Nanoparticles*

The nanoparticles used in clinical applications needed surface optimization. In a drug delivery system for better drug delivery potential, it is beneficial to functionalize and coat the nanoparticles. The selection of coating material is very important in biomedical applications because surface modification sometimes changes the nanoparticle's properties and also their performance in biomedical applications.

Magnetic nanoparticles (MNPs) are used in different biological applications like magnetic hyperthermia induction in tumors, magnetic resonance imaging (MRI), and diagnostic imaging. Magnetic nanoparticles (MNPs) used for the above-mentioned applications do not agglomerate due to magnetic or colloidal in the specific medium. To avoid these problems coated nanoparticles can be used to facilitate solubilization in the provided medium. Biopolymers are suitable materials for these requirements due to their biodegradability and bioactivity.

9.3 Application

Different polymer coatings are used for different applications like Poly(2-methyl-2-oxazoline) (PMOXA) coatings are used for stainless steel bioactivity, antifouling properties, preventing late stent thrombosis and in-stent restenosis. Polytetrafluoroethylene PEO coatings are used for protective and antifriction properties.

Hydroxy-apatite-polytetrafluoroethylene coatings for corrosion resistance and impart bioactivity PVDF Spin coatings are used for grown film wearable and a wireless pressure sensor for heart rate monitoring. Polyaniline-coated with PVDF is used for human health monitoring.

10 Conclusion

Smart polymers and composites are versatile in applications. Their use in different fields is increasing day by day because of technological advancements. Smart polymers used as biopolymers reduce CO_2 and harmful chemicals emission and degrade easily without affecting the environment. These polymers are also used as adsorbents for cleaning water from dyes. Smart polymer composite in the fuel cell has provided the solution for reducing greenhouse gas emissions, highly reliable, and flexible to install. They are a better option than conventional fuel cells. Polymeric membrane technology has opened new horizons to provide the efficient application. Recent developments in nanocomposite membranes are stimuli response factors that made them much more effective than conventional membranes. The use of polymer is not limited to this application; they found their ways to medical science, tissue engineering, drug delivery, and bio-degradable bandages. In industry bio-compatible, bio-degradable polymers have been used for active packaging to store food for a period until they start degradation and show a smart display for the consumers under the expiry of the food item.

References

1. Aili, D., Jensen, J.O., Li, O.: Polymers for Fuel Cells Fluorinated Polymers Poly(Perfluorosulfonic Acid), pp. 1–13 (2014). https://doi.org/10.1007/978-3-642-36199-9
2. Kim, Y., Yoo, S., Kim, J.H.: Water-based highly stretchable PEDOT: PSS/nonionic WPU transparent electrode. Polymers (Basel) **14**(5) (2022). https://doi.org/10.3390/polym14050949
3. Namkoong, G., Younes, E.M., Abdel-Fattah, T.M., El-Maghraby, E.M., Elsayed, A.H., Abo Elazm, A.H.: Aging process of PEDOT: PSS dispersion and robust recovery of aged PEDOT: PSS as a hole transport layer for organic solar cells. Org. Electron. **25**, 237–244 (2015). https://doi.org/10.1016/j.orgel.2015.06.049
4. Rashid, I.A., et al.: Stretchable strain sensors based on polyaniline/thermoplastic polyurethane blends. Polym. Bull. 0123456789 (2019). https://doi.org/10.1007/s00289-019-02796-x
5. Zahid, M., et al.: Recent developments in textile based polymeric smart sensor for human health monitoring: a review. Arab. J. Chem. **15**(1), 103480 (2022). https://doi.org/10.1016/j.arabjc.2021.103480
6. Rashid, I.A., et al.: Electrically conductive epoxy/polyaniline composite fabrication and characterization for electronic applications. J. Reinf. Plast. Compos. **0**(0), 07316844211023991. https://doi.org/10.1177/07316844211023991
7. Shakir, H.M.F., Tariq, A., Afzal, A., Abdul Rashid, I.: Mechanical, thermal and EMI shielding study of electrically conductive polymeric hybrid nano-composites. J. Mater. Sci. Mater. Electron. **30**(18), 17382–17392 (2019). https://doi.org/10.1007/s10854-019-02088-0
8. Shakir, M.F., et al.: EMI Shielding Characteristics of Electrically Conductive Polymer Blends of PS/PANI in Microwave and IR Region (2019). https://doi.org/10.1007/s11664-019-07631-7
9. Shakir, M.F., et al.: Effect of Nickel-spinal-Ferrites on EMI shielding properties of polystyrene/polyaniline blend. SN Appl. Sci. **2**(4) (2020). https://doi.org/10.1007/s42452-020-2535-4
10. Moradi, O., Sharma, G.: Emerging novel polymeric adsorbents for removing dyes from wastewater: a comprehensive review and comparison with other adsorbents. Environ. Res. **201**(January), 111534 (2021). https://doi.org/10.1016/j.envres.2021.111534
11. Aili, D., Jensen, J.O., Li, Q.: Encyclopedia of polymeric nanomaterials. Encycl. Polym. Nanomater. 1–13 (2020). https://doi.org/10.1007/978-3-642-36199-9
12. Ryu, T., et al.: Synthesis and characterization of polymer electrolyte membrane containing methylisatin moiety by polyhydroalkylation for fuel cell. Int. J. Hydrogen Energy **43**(10), 5398–5404 (2018). https://doi.org/10.1016/j.ijhydene.2017.12.164
13. Schenk, A., Cermenek, B., Hacker, V.: Other Polymer Electrolyte Fuel Cells. Elsevier Inc. (2018)
14. Siekierka, A., Smolińska-Kempisty, K., Wolska, J.: Enhanced specific mechanism of separation by polymeric membrane modification—a short review. Membranes (Basel) **11**(12) (2021). https://doi.org/10.3390/membranes11120942
15. Ray, S.S., Iroegbu, A.O.C., Bordado, J.C.: Polymer-based membranes and composites for safe, potable, and usable water: a survey of recent advances. Chem. Afr. **3**(3), 593–608 (2020). https://doi.org/10.1007/s42250-020-00166-z
16. El-Ghoul, Y., Alminderej, F.M., Alsubaie, F.M., Alrasheed, R., Almousa, N.H.: Recent advances in functional polymer materials for energy, water, and biomedical applications: a review. Polymers (Basel) **13**(24) (2021). https://doi.org/10.3390/polym13244327
17. Othman, S.H.: Bio-nanocomposite materials for food packaging applications: types of biopolymer and nano-sized filler. Agric. Agric. Sci. Procedia **2**, 296–303 (2014). https://doi.org/10.1016/j.aaspro.2014.11.042
18. Basavegowda, N., Baek, K.H.: Advances in functional biopolymer-based nanocomposites for active food packaging applications. Polymers (Basel) **13**(23) (2021). https://doi.org/10.3390/polym13234198
19. Ulbrich, K., Holá, K., Šubr, V., Bakandritsos, A., Tuček, J., Zbořil, R.: Targeted drug delivery with polymers and magnetic nanoparticles: covalent and noncovalent approaches, release

control, and clinical studies. Chem. Rev. **116**(9), 5338–5431 (2016). https://doi.org/10.1021/acs.chemrev.5b00589
20. Murthy, N., Campbell, J., Fausto, N., Hoffman, A.S., Stayton, P.S.: Bioinspired pH-responsive polymers for the intracellular delivery of biomolecular drugs. Bioconjug. Chem. **14**(2), 412–419 (2003). https://doi.org/10.1021/bc020056d
21. Nathanael, A.J.: Biopolymer Coatings for Biomedical Applications (2020)
22. Heyse, P., De Vilder, I., Vanneste, M.: 4—Smart Durable and Self-Healing Textile Coatings. Elsevier Ltd (2016)

Self-healing Elastomers

Zakariya Zubair, Ahmad Usman, and Asif Hafeez

Abstract The research in the field of self-healing elastomers (SHE) has resulted in different complex structures with various healing strategies (hydrogen bonds, covalent and non-covalent interactions and combinations of both). Elastomers with good mechanical performance and healing efficiency typically do not exist, as self-healing mechanism limits the mechanical properties. To get optimum mechanical and healing properties, using a combination of different mechanisms is emerging as a possible solution in SHEs. This chapter, mainly focused on SHEs, in which covalent bonds in combination with non-covalent interactions provide the best balance between repairability and mechanical performance. The applications of SHE include sensors, controlled drug release, coatings, actuators, railway components, hoses, seals, gaskets, and tires used in high-performance applications.

Keywords Self-healing · Elastomers · Hydrogen bonding · Thermoplastics · Covalent bonds · Self-sealing

1 Introduction

The last 100 years have witnessed amazing breakthroughs, resulting in the production of new materials with the desired characteristics [1]. Self-healing (SH) has been an inevitable trend among these properties over the past 20 years. Keeping in view the sustainability, the SH materials are the most desirable as it increases product life span, causing decrease in environmental waste. The circular economy presents challenges for some polymers, such as elastomers and thermosets, which are difficult to reprocess despite substantial efforts [2]. But due to their significant demand for industrial applications, it is crucial to make them appropriate for the circular economy paradigm and inducing SH characteristic is one approach to accomplish this [3].

Z. Zubair (✉) · A. Usman · A. Hafeez
Department of Materials, School of Engineering and Technology, National Textile University, Faisalabad, Pakistan
e-mail: zzubair@ntu.edu.pk

K. Shaker and A. Hafeez (eds.), *Advanced Functional Polymers*, Engineering Materials,
https://doi.org/10.1007/978-981-99-0787-8_6

1.1 Natural Self-healing Phenomenon

Human skin is by far the most known example of how tiny injuries to the skin heal completely but more severe injuries leave scars. Both situations almost completely restore the system's operational potential [4]. Other organisms' capacity for repair can be considerably more surprising; for instance, the Cnidarian hydra can reconstruct its entire body from its broken-up cells. The Lerna Hydra was viewed as the pinnacle of SH in Greek mythology [5].

Although the development of synthetic SH materials has long been a dream, there has been a notable rise in optimism in this area over the past 10 years. Polymers fall in two different types i.e. thermoplastics and thermosets, which are being used in several high-tech applications [6, 7]. To regain their mechanical properties, SH polymers need to be coupled across the interfaces created by rupture, breaking, or cracking. The crosslinks prevent polymer chains from creating entanglements, limiting the self healing in thermosets [8].

Reversible covalent and composite are the two main approaches that have emerged for SH in thermoset materials. When an external trigger like light or temperature is applied, covalent bonds are rearranged in the first method, which is frequently referred to as "self-mending": fractures in the material are fused by newly generated bonding, as in a reversible Diels–Alder reaction studied by Fred Wudl's research team (Fig. 1a). It is clear that this method has a lot of potential for reversible (but powerful) non-covalent bonds, and efforts to use hydrogen bonds for this purpose are being researched.

The second method, created by White, Sottos, and others, involves embedding monomers carrying microcapsules within the thermoset material. When a fracture appears in the capsules' skin, the capsules burst, and the catalyst is dispersed throughout the matrix, causing the unreacted monomer to discharge and harden (Fig. 1b). Since no external trigger is required to initiate healing, this sort of SH is referred to as "autonomous" [9]. Thermoplastics naturally have simpler SH than thermosets because heating induces chain mobility, which can result in the formation of entanglements across surfaces. Thermoplastic interfaces can be joined together without the need for an additional adhesive, unlike thermosets. Sometimes the material's impact, which can harm the substance, can supply the temperature needed to produce movement. This procedure is known as "ballistic impact self-repair" [10].

1.2 Self-healing Mechanism in Nature

A quick self-sealing (SS) leaves residual damage (fracture, cut, or crack) that has been repaired well to the point where it can continue to function as before (stiffness, self-cleaning), but the mechanical properties (rigidity, hardness) are not recovered [11]. The biological models of animals and plants have come to reflect hierarchically

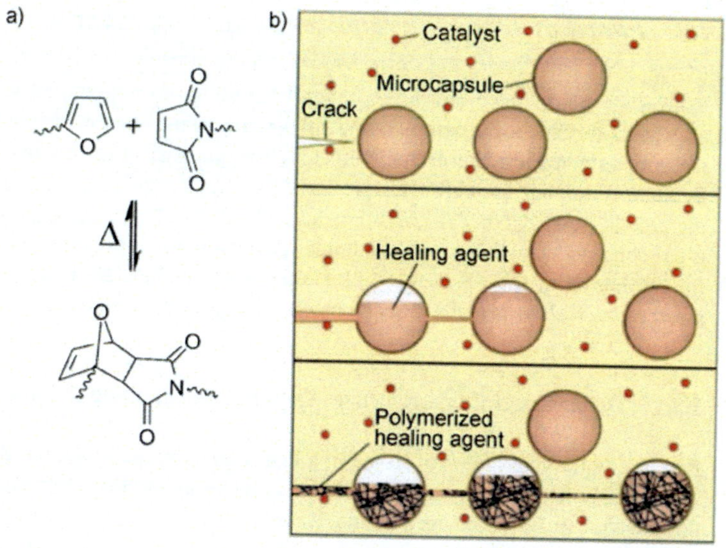

Fig. 1 Hitherto explored two different primary approaches; **a** reversible-covalent bonds and **b** thermoset matrix, embedded with microcapsules [10]

structured material systems. The number of hierarchical levels, their varying architectures and components, and the changes in functional requirements demonstrate how animals and plants developed a variety of SH strategies. All materials have a preliminary SS stage that quickly seals a wound or injury, which can be distinguished from a succeeding SH stage that leads to the most accurate possible recovery of the initial condition, despite different periods and procedures [12] (Fig. 2).

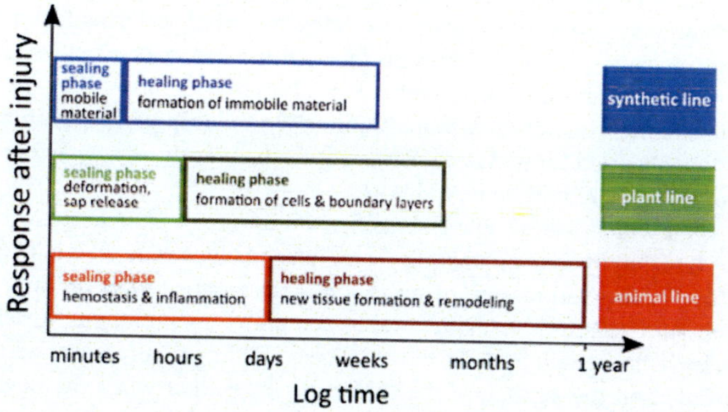

Fig. 2 Comparative analysis of self-repairing stages in technical materials, animals and plants [12]

1.3 *Plants as Idea Generators or Inspiration Sources*

With time, plants have developed several variations to dehydrated locations, such as water storage in the parenchyma of the succulent plant body and a reduction in branching, which causes moisture loss via quick sealing and bruising healing. Because quick and synchronized wound closure is essential for the development of the plant's water balance in conditions of water stress, the encapsulation mechanism is safe guarded several times by redundant mechanisms. The primary driving factors for the leaves of the SS in the Pink Carpet plant include hydraulic shrinkage and swelling as well as growth-induced mechanical prestresses [13].

1.3.1 Delosperma Cooperi Leaves-Inspired Self-healing Polymer

By forcing the five layers of tissue in a leaf to expand and contract simultaneously, a wound creates a mechanical imbalance that causes the leaf to change form until a new mechanical equilibrium is found. Based on a mechanically driven SS mechanism previously found in D. cooperi succulent leaves, Yang et al. developed a bioinspired SH polymer. They developed a commercial microphase detachable copolymer based on a biological function model composed of separate tissues with intrinsic strain and stress that has shape memory capabilities that enable the SH following mechanical injury [14].

1.3.2 Latex-Bearing Plants-Inspired Self-healing Elastomers

The weeping Fig. 3 (Ficus benjamina), the rubber tree (Hevea brasiliensis), and the various spurge species (Euphorbia spp.) were selected as the best case studies to address the technological challenge of limiting fracture propagation in self-repairing material systems. To examine how latex coagulation during fast fissure SH affected the recovery of tensile strength in F. benjamina bark samples, Bauer and Speck utilized mechanical experiments [15]. Figure 3 demonstrates that tensile strength substantially increases within 30 min of external injury, accounting for 55% of the value obtained for undamaged bark. If the latex is thrown away before the time limit runs out, there will be no SH [12].

It is widely known that H. brasiliensis coagulates plant latex. Microtubes are made of latex emulsions containing membrane vesicles (laticifers). Within the laticifers, there is an overpressure of up to 1.5 megapascals. After an injury, the emulsion's pressure decreases, causing protein-containing membranous vesicles to rupture and release the hevein proteins they had been holding in. protein-binding sites which are located on the surface of the rubber particles, and when the hevein protein forms dimers with these protein-binding sites (beginning with Ca^{2+} ions), latex particle coagulation results [16]. Innovative bioinspired SHEs with expressive mechanical

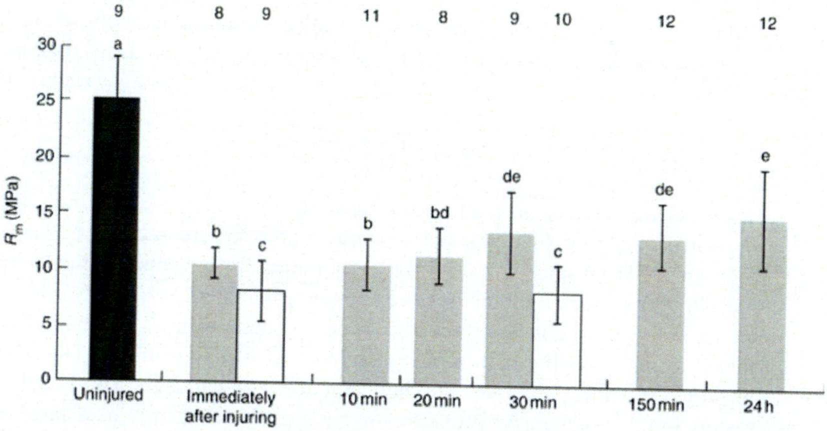

Fig. 3 SH and SS phenomenon in the barks of Ficus benjamina tree. Rm (tensile strength) of unmarked bark (the black columns), and at different times after mark (gray columns). The white columns show bark specimens from which the latex was immediately wiped after mark. Results marked with same letter don't differ significantly [12]

restoration following a macroscopic incision were created using two ways. These SHEs were dependent on the SH mechanism.

First, SH ionomeric elastomers with self-repair capabilities during the nonvulcanized stage but not after full vulcanization were created based on the function of Ca^{2+} ions. Chemically mimicked microcapsules containing SH agents without a hull were created using microphase detached NBR (nitrile butadiene rubber) or hyperbranched PEI (polyethyleneimine) blends in a second method [17].

1.3.3 Nacre-Inspired Self-healing Composites

Because it creates an exceptional mechanical characteristic in combination with strength, toughness, and stiffness, the brick and howitzer edifice of a rigid inorganic portion (95 volume percent interlocking aragonite platelets) and a comfortable biological matrix (5 volume percent protein-polysaccharide) in nacre have attracted a lot of attention in recent years. According to fracture morphologies, Feng et al. found that the three main toughening mechanisms operating on nacre are crack deflection, fiber pullout, and organic-matrix bridging [18]. The remarkable properties of the nacre, a substance made by some mollusks, have inspired many scientific organizations to create synthetic materials with properties similar to the nacre. These attributes include the combination of toughness and stiffness that leads to long-lasting structural ceramics. By combining dynamic polymers with low glass transition temperatures, four hydrogen-bonding motifs, and high aspect-ratio synthetic nano-clays, Zhu et al. were able to produce films that resembled nacre. Because of their exceptional dynamics, all so-called EG-UPy-()-polymers self-heal swiftly and independently following a rupture or cut, and when combined into nanocomposites inspired by nacre, they at least provide the healing of nanoscopic and nanovoid fractures [19].

1.3.4 Hemostasis-Inspired Self-healing Composites

Typically challenging to diagnose, structural issues like delamination, debonding, or matrix cracking finally result in catastrophic failure scenarios. Among other measures, SH systems can stop these problems. In response to these technological difficulties, SH methods for the lightweight materials listed below have been created [20]. Pang and Bond created "bleeding composites," which are SH and are meant to resemble the natural mending process in living things. The release of a healing agent that penetrates the damaged area and can significantly subsidize the restoration of mechanical properties as well as the clemency of a fluorescent tint that contributes to the ocular improvement of the bruise section are two functional principles of the bleeding behavior of liquid-filled cavern fibers. In this instance, hollow glass tubes that function as plies in the composite are packed with healing components (resin with hardener) and fluorescent dye. Based on the healing analysis and healing therapy, these composite systems show point-bending flexural testing [21].

1.4 *Self-healing Mechanism in Polymers*

When SH occurs in polymers that mechanical characteristic must be recovered by fixing across interfaces created by cutting, breaking, or cracking. It is particularly difficult to repair in thermosets because crosslinks prevent polymer chains from tying surfaces together by creating entanglements over the boundary.

1.4.1 In Thermoplastic

Thermoplastic elastomers are an effective and valuable family of thermoplastics with flexible (rubber-like) properties and essentially no viscous flow at operating temperatures because the chains form physical crosslinks in glassy or crystalline zones with restricted mobility. SH requires a high degree of chain motility without the aid of cures or external stimuli, which seems to conflict with the fixing needed by thermoplastic elastomers to form a lasting network. This remarkable fusion of elastic and SH properties was discovered earlier this year [22]. Leibler and colleagues discovered the complete and spontaneous restoration of mechanical properties in a thermoplastic elastomer by only bringing broken surfaces together.

The system is distinguished by endlessly repeated healing at a low cost and with simple processing, suggesting that it is likely to be applied in industry, as well as thermoplastic melt behavior and consequently less toxicity. This uncommon confluence

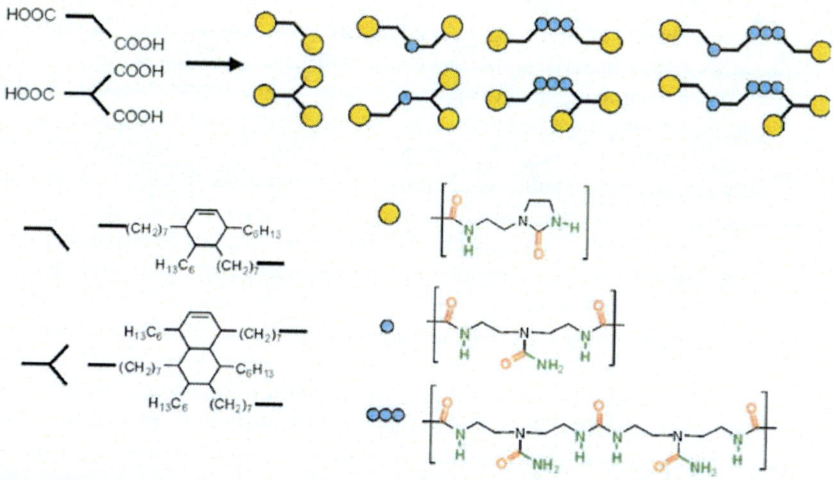

Fig. 4 Schematic diagram of Leibler's SH rubber [10]

of characteristics marks a turning point in the understanding of the SH mechanism and explains why this discovery has attracted the interest of the scientific community, popular science media, and the international press. Leibler's method consists of a few straightforward stages. Diethylenetriamine and later urea are used to treat dimer and trimer fatty acids that are derived from renewable herbal oils (Fig. 4).

The result is a transparent vitreous solid with a 288 °C Tg (glass transition temperature). Multiple di- and trifunctional structural building blocks are combined with various urea and amide compounds that have strong hydrogen bonds to form a supramolecular connection. Because there is a wide variety of species, crystallization is prevented, resulting in higher degrees of stretchability. High temperatures result in a significant reduction in viscosity due to the temperature-dependent strength of hydrogen-bonding units, which makes it possible to quickly melt-process the material into any desired shape [23]. The glass transition temperature (Tg) was decreased, converting the substance into a thermoplastic elastomer by using dodecane (11% by weight) as a plasticizer. Tensile tests revealed elongations at break of more than 500%, which is typical of soft rubber.

Rheological measurements in the shear mode, which showed normal lifetimes for the network of many weeks, revealed the truth of rubbery behavior. This is essential for the material to have low creep: the sample (material) had low and no creep when it was subjected to initial strains of 144 and 32%, respectively, over a long period under continuous loads. Additionally, it was shown that creep is almost entirely recoverable, which means that the sample gradually takes on its former shape. This essential requirement for a functional rubbery material is adequately met by Leibler's substance. The low but significant creep reported here is not conclusive, although the majority of conventional rubbers have no creep at all; the material is likely to work satisfactorily if the right operating circumstances are met. The most surprising quality of this material is its capacity for self-healing [24].

When a sample is divided into two halves by cutting or shredding it, the halves can be brought together and squeezed together to restore the sample's original strength. While bonding takes time to begin and only takes 3 h to attain its maximum strength, adhesion almost immediately takes hold and is strong enough for elements to support their weight. How do Leibler and his friends prevent too much creep from endangering SH? Understanding SH in this material necessitates having a grasp of the kinetics and concentration of highly coupled hydrogen-bonding groups. The hydrogen bonds, not the covalent bonds, are destroyed when the elastomer ruptures, leaving a substantial layer of unconnected groups at the contact. Non-connected hydrogen bonds persist for a while, looking for a friend when they come into contact with another freshly cut surface. The free hydrogen-bonding units on the first surface will automatically look for alternate companions within the same piece if the second surface is not available. The healing process was significantly slower and required longer healing durations when shattered portions were separated for 6 or 18 h before being pressed together. The relatively quick group rearrangements at the boundary produce a special mending process that joins chains that are even a part of the networks on both sides of the boundary rather than relying on the creeping motion of entire polymer chains to restore mechanical properties. In this way, the procedure is practically comparable to Wudl's "reversible-covalent" method [10].

1.4.2 In Elastomers

SH is a valuable characteristic of living tissues that enables them to appropriately mend themselves after suffering mechanical harm. This SH feature, which was inspired by nature, increases the stability and durability of artificial materials, reduces maintenance expenses, and opens up new applications. As a result, SH materials have drawn a lot of attention and have demonstrated great promise in many industries, including flexible robotics, electronic enclosures, and automotive coatings. The first SH compounds in this field 20 years ago were based on extrinsic mending reagents. Their ability to heal is constrained due to a lack of healing agents. Because of this, current research is concentrating on naturally healing substances that rely on non-covalent interactions or reversible-covalent bonds. Inputs of external energy, such as heat and light, are typically needed for the SH process. The development of materials that can self-heal at 25 °C is particularly beneficial because many materials deteriorate in real life when there are no external stimuli present [25].

A typical method for creating materials that heal at room temperature involves taking into account non-covalent interactions such as metal–ligand coordination, hydrogen bonds, host–guest interactions, ionic interactions, and physical forces. A limited number of non-covalent contacts may enhance SH. The materials that are created are, however, typically somewhat fragile. However, a high number of non-covalent connections may compromise the material's SH, flexibility, and stiffness while enhancing mechanical properties. Due to their linear molecular structures, these materials may also exhibit low resistance and probable creep. Many scientists have used dynamic covalent contacts including urea bonds, boron–oxygen bonds, and ionic bonding to generate crosslinked structures to produce reasonably robust healable materials. Conversely, crosslinked networks hinder chain movement and reduce healing capacity. Usually, a material's ability to self-heal and its mechanical properties are interdependently special. Due to their occasionally contradictory demands on molecular structures, it is currently challenging to simultaneously achieve high mechanical robustness and healing efficiency, especially in ambient conditions [26].

Successfully developed SHEs have excellent mechanical qualities and potential functionality. However, although this additional function can significantly expand their uses, adhesion force has gotten relatively little research in these SHEs. SHEs with good stickiness are crucial in applications including artificial skins, wearable electronics, and sealants where direct contact of the functional moieties with elastic substrates is required throughout the lifespan of the product. For instance, the wearable strain sensors described by Zeng et al. depend on close, sustainable contact with human skin to accurately detect human activity, and a break from the skin could impair functioning. In addition, the SH capability extends the lifespan of the device by automatically healing any fractures that may develop over time. Several elastomers simultaneously exhibit SH and sticky properties. However, the bulk of them has significant flaws, such as weak mechanical performance, restricted adhesion strength, or strong adhesion only to specific substrates [27].

2 Evolution of Self-healing Polymers

The progression of SH polymers has occurred in multiple SH strategies that have yielded complex structures competent for supporting numerous cycles, among other properties. This advancement enables us to offer the exertion of a timetable that segregates generations of SH polymers based on the healing mechanism and associated with historical development.

2.1 Key Concepts in Self-healing

Influenced by natural phenomena, SH substances possess the potential to self-heal or self-repair harmful imitating mechanisms originating in living things, like humans and plants. Three ideas have been developed to assure the success of SH: (1) localization; (2) temporality; (3) mobility, as well as a fourth crucial idea; and (4) a classification system (shown in Fig. 5) for different groups of SH materials. The term "localization" is related to the location or magnitude of injury in a product. It might exist on the surface, like scratch marks, (micro) fractures, or cuts, or it might be inside the product, such as the extension of fiber bonding, surface damage, or delamination, leading to serious breakdown or molecular scale damage, such as material network rupture. When examining the material's SH potential, the location and magnitude of these damages are important. The goal is to develop a single technique that ensures healing on all scales. The second component, for the time being, is determined by the difference between the time of occurrence of the injury and its correction. Self-healing is a slow process, not immediate, even in nature. The goal is to decrease the time it takes to heal the body. The third important principle, giving mobility to the material, is one approach to shortening such time. Mobility helps in the delivery of the healers to the damaged region and reestablishment of broken bonds. This concept is important for helping others; for example, if the agent's mobility is insufficient, it will not flow toward the injury or will flow slowly [28].

The mechanism is the final crucial concept in SH. This concept can be used to differentiate between extrinsic and intrinsic SH materials. Extrinsic SH materials depend on an external agent to initiate the healing process, typically in the form of capsules or circulatory networks. Such agents are launched to repair the damage but have no interaction with the matrix. Intrinsically SH materials, on the other hand, are those in which the material's reversible bonds may be reestablished following a damaging event. Extrinsic systems are utilized in thermosets, primarily epoxy-resins, but intrinsic systems are heavily studied in elastomers including silicones, polyurethanes, and GP rubbers [29]. On the basis of SH phenomena and evolution, the SH substances are classified into four generations (Fig. 6).

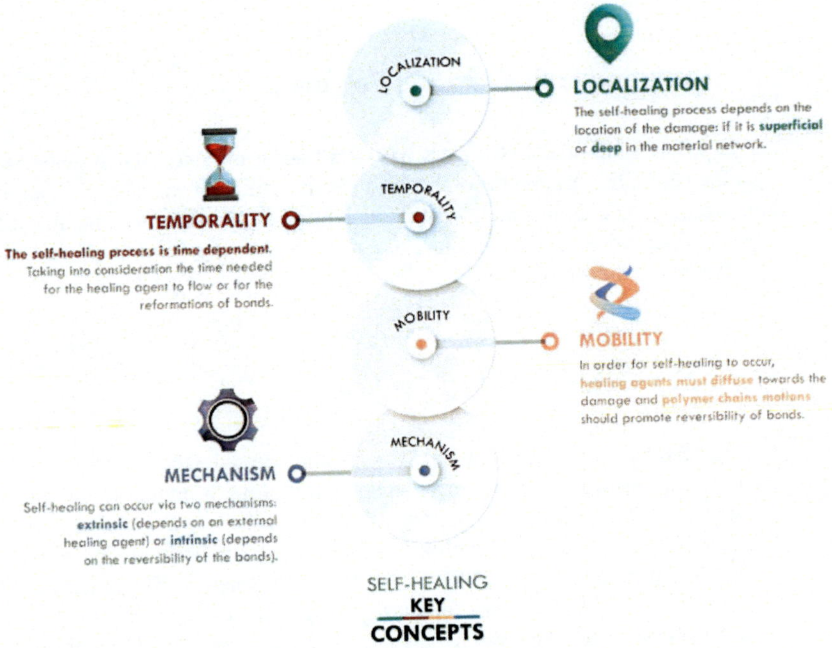

Fig. 5 Key concepts in SH [28]

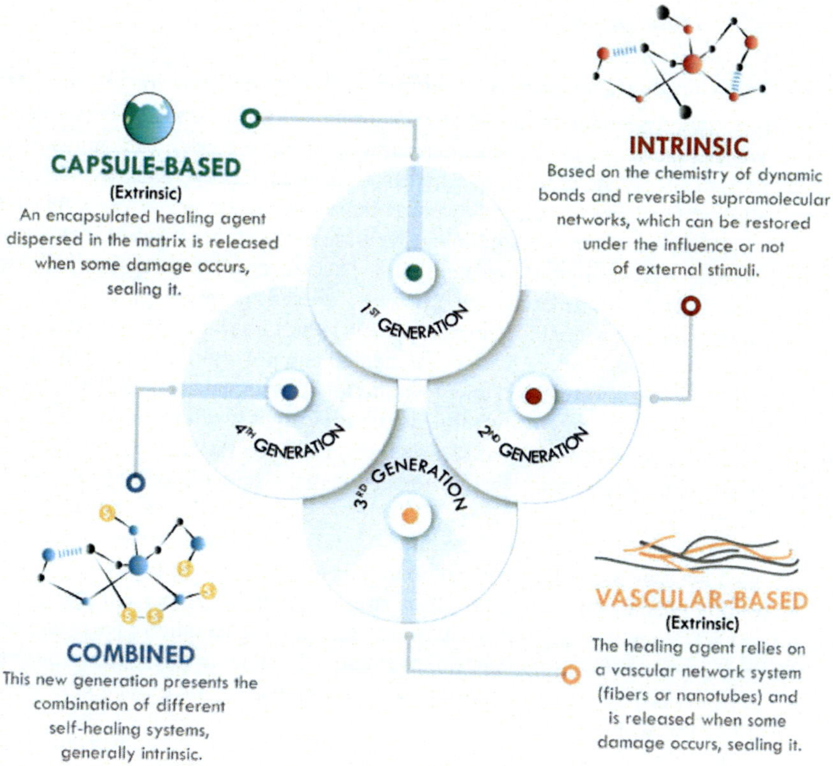

Fig. 6 The SH mechanism involved the different generations of SH materials [28]

Extrinsic processes were used in the first generation of SH substances, which used encapsulated external healing substances. This generation has the limitation of being able to support a single SH cycle only. To address this limitation, the second generation of SH polymer-based substances was developed, based on intrinsic processes and reversible bonds. The material's SH ability and strength, on the other hand, were incompatible: enhancing one meant reducing the other. The intrinsic concept has been investigated in a wide range of polymers, with a focus on elastomers. Furthermore, extrinsic mechanisms were further developed by enclosing healing in vascular networks, opening the pathway for the third generation of SH polymer-based substances. The third generation of SH polymers was inspired by nature. Finally, the fourth generation is rapidly developing, intending to overcome the flaws of the prior generations. As a result, the goal is to combine several healing processes to create a polymer having outstanding mechanical characteristics, good healing efficiency, and resilience to numerous injury loops [30].

2.2 Timeline of Self-healing Polymers

Malinskii et al. published some of the early studies on polymeric SH, notably in PVA (poly-vinyl acetate), in the 1970s, according to the literature (PVAc). Jud et al. and Wool et al. later conducted an extensive study on self-healing in polystyrene (PS), poly-methacrylate (PMA), and hydroxy-terminated polybutadiene (PB). However, all of these investigations, as well as those conducted in the years that followed, were built on chain crosslinking, a well-known polymer concept that requires just a temperature marginally greater than the substance's glass transition (Tg) to occur. According to White et al., the SH was achieved by mixing a healing agent dicyclopentadiene contained in microcapsules with a Grubb's catalyst (platinum catalyst) spread in resin (epoxy). The dicyclopentadiene polymerizes and plugs the fracture when the agent is released and gets into touch with the catalyst. This technique, in its early phases, may recover the maximum weight in a toughness test at up to 75% effectiveness. Keller et al. used a chemical based on the two kinds of microcapsules to implement this method in elastomers, namely, poly (dimethylsiloxane) (PDMS). They entrapped a copolymer (PDMS) and an initiator with an active site that serves to associate the vinyl groups of the bifunctional resin through platinum catalyst accomplishment in one. This chemical process, which also depends on the polymerization of other chemicals, allowed for tears strength of up to 120 percent efficient recovery [31]. The healing agents and catalysts in this initial generation have been categorized in many ways based on their arrangement. The most frequent categories are single capsule, disperse catalyst/capsule, process capsules/droplet, and double capsules, so all capsules (Fig. 7).

In the second generation, dynamic bonding chemistry is applied. Dynamic bonds and contacts are those that are changeable under equilibria and might be covalent or non-covalent. Although reverse chemistry-based polymers have previously been

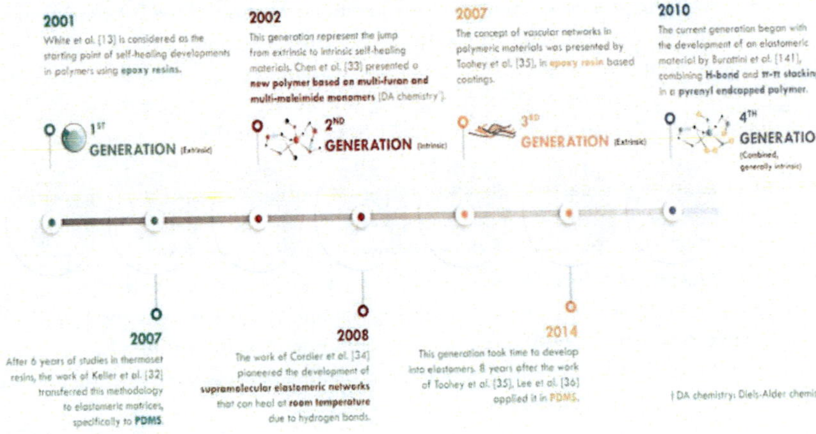

Fig. 7 Different generations of SH polymers [28]

established, Chen et al. created the first SH polymers employing multi-maleimide and multi-furan monomer (Diels–Alder chemistry). Years ago, Cordier et al. discovered and manufactured molecules capable of generating chains and linkers by hydrogen bonding (HB), for the first time incorporating the inherent SH process into an elastomer. As a consequence, they built a supramolecular association capable of self-restoration at ambient temperature [32].

Toohey and colleagues created the third phase of SH polymers. The ultimate stimulation for this generation took a long time to design due to the complexity of combining vascular systems into a polymer matrix. Third generation is categorized based on the properties of the vascular network and the technology used to prepare it. A fourth generation is evolving in this sector. This new phenomenon does not, however, imply that previous approaches are no longer being researched.

2.3 *Fourth Generation of Self-healing Elastomers*

Because of the research of Burattini et al., the scholarly works on merged SH processes have been growing steadily and it has centered on SH intrinsically methodologies, every time searching for an optimum solution pairing of bonds dynamically, whether covalent (the ones that require stimulation for reversibility) or non-covalent (those that do not require a stimulus for reversibility) (due to their ability those are reversible intrinsically) [32].

2.3.1 Combined Non-covalent Systems

In intrinsic SH processes of non-covalent nature, all poor interactions which can occur among various families of atoms, like van der Waals forces, ionic interactions, HB, stacking, metal–ligand cooperation, dipole–dipole interrelations, and presenter interactions, are included. According to some researchers, the presence of the effect of shape memory is responsible for SH abilities (SM). It's controversial if SM may be considered a process of SH in itself. All non-covalent interactions are summarized in Fig. 8 along with their fundamental description [33].

Although the bonding energy of non-covalent systems is lower than that of pure covalent systems, this system offers better efficiencies of healing because they make it easier to restore broken connections even though kept at room temperature. HB is the most common interaction of non-covalent in elastomers, which has been paired with several other interactions of non-covalent to generate materials for SH with a range of characteristics. Burattini et al. presented a combination of healing components (π-π-stacking and h bonding) in an elastomer network based upon polyimide and PU (polyurethane) including pyrenyl end groups.

The stacking of π-π was because of the diamide groups' π-electron deficiency and the pyrenyl units' π-electron abundance. Meanwhile, hydrogen bonds are formed between the pyrenyl groups' terminal residues in polyurethane at the intermolecular

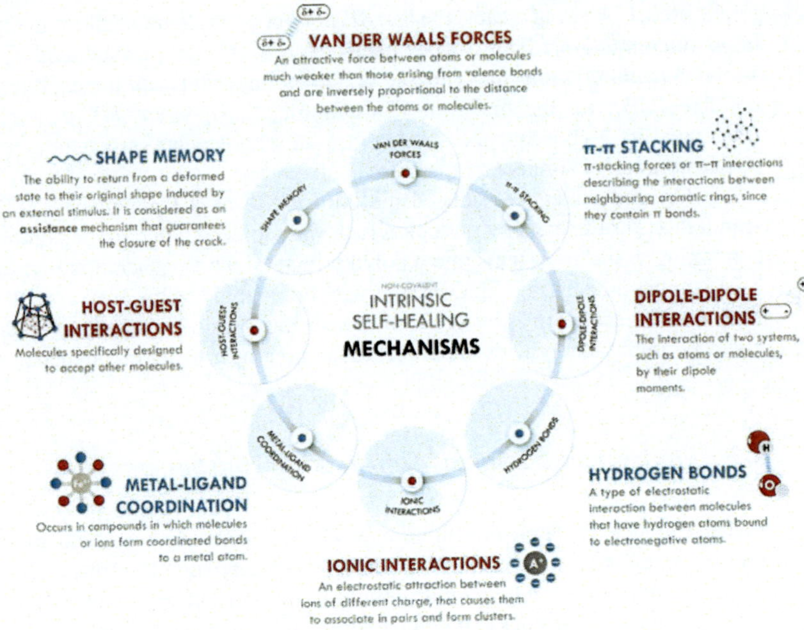

Fig. 8 Intrinsic non-covalent SH mechanisms [28]

level. 0.2 MPa was the tensile strength of this material with the efficiency of healing at 80% or more for tensile strength. Ionic interactions are the second most frequent non-covalent mechanism. Xu et al. successfully coupled chitosan nanoparticles with HB to carboxyl styrene-butadiene rubber (XSBR) stacked with them. The synthesis of ionic clusters at room temperature resulted in the network formation of crosslinked supramolecular with reversible links and healing efficiency of up to 92 percent and 1.3 MPa tensile strength. For the first time, Sattar et al. utilized the same combination to make silica-loaded (SiO_2) natural rubber (NR) composites. In their process, ionization of the elastomer's inherent proteins and lipids was employed to form ionic crosslinks [33]. The dynamical supramolecular network was established by the addition of magnesium sulfate ($MgSO_4$), which formed Mg^{2+} ions that produced electrostatic interactions with the negatively charged lipids as a result of acidic ionization. At 50 °C, this procedure yielded healing efficiency of 79% and 18.5 MPa of tensile strength. Fourier-transform infrared spectroscopy (FTIR) compared with hydrolyzed NR indicated the creation of the ionic network, which had a healing efficiency of just 52%.

Polyurethane (PU) was also shown to have these two mechanisms. Because PU and its alternatives are the materials that were studied the most in terms of the mechanism of the intrinsic process, it's not astonishing that they were the first to report a network comprised of three kinds of dynamic bonds: one covalent and two non-covalent, the closing of which is in charge for SH at ambient temperature. Zhang et al. developed a polyurethane elastomer based on a Cu (II)-dimethylglyoxime-urethane complex (Cu-DOU-CPU). Copper ions provided coordination in metal–ligand and formed hydrogen bonds between groups of amino and ester's main chain. Researchers achieved 14.8 MPa tensile strength and 92 percent efficiency, which could be used in wires [34].

2.3.2 Combined Covalent Systems

Intrinsic covalent processes include any chemical bonds that form between other atoms and can be active in reaction to an external incitement. A schematic representation of several of these bonds as well as their fundamental description is shown in Fig. 9. Disulfides, which can experience catalyzed processes, and Diels–Alder chemistry, in which retro-Diels–Alder and Diels–Alder interactions occur at various temperatures, are other notable examples. Since these bonds have higher energy over non-covalent interactions, their participation is significantly connected to the mechanical performance of the material. Further, it is an important aspect of obtaining high healing efficiency [35].

In opposition to aliphatic disulfides, which require external stimuli, Lee et al. used aromatic disulfides, which may undergo metathesis at room temperature [36]. They developed a PDMS that could be repaired in 4 h, but its mechanical properties

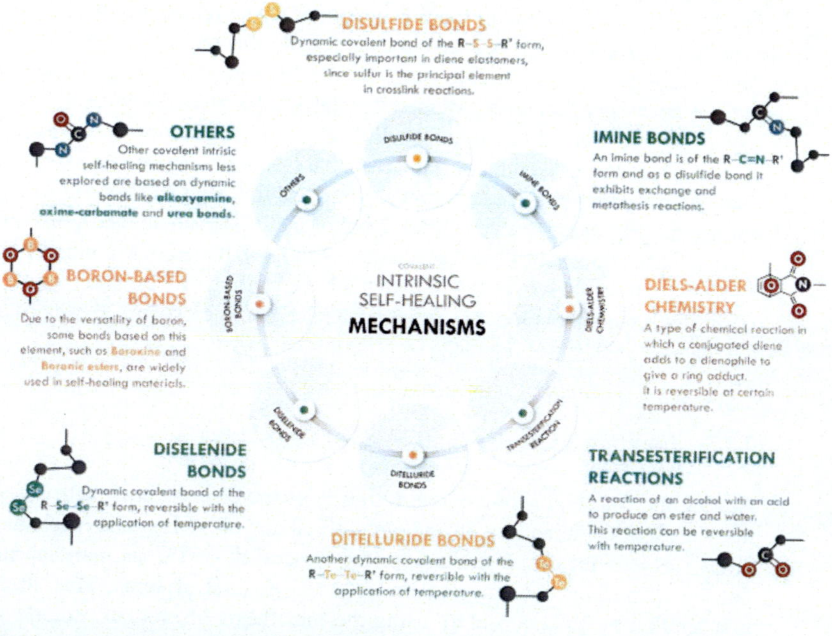

Fig. 9 Intrinsic covalent SH mechanisms [28]

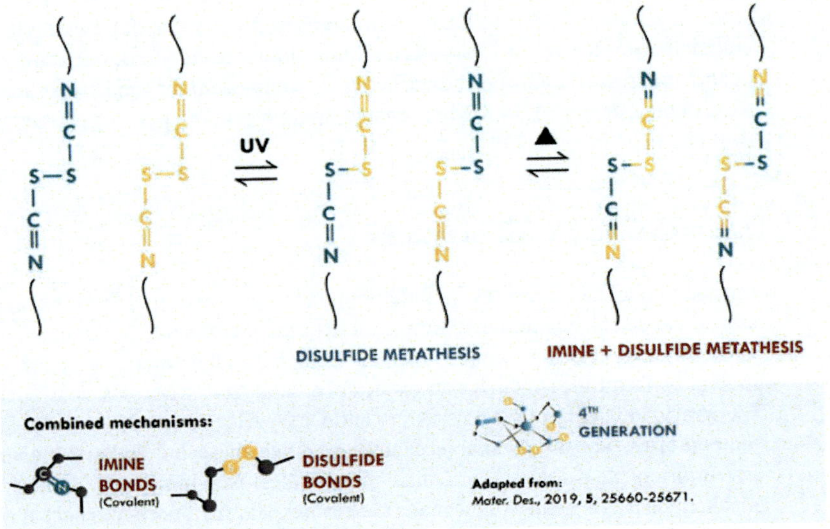

Fig. 10 Disulfide and imine metathesis in PU [28]

were only 0.15 MPa, which is typical of siloxane-based systems. The disulfide bond, which acts as a sacrificial bonding, is primarily responsible for SH, while the imine bond serves as a crosslink of the semi-permanent site for flexibility and keeping the actual form. Despite its low mechanical strength, this research provides a prototype for some applications (such as adhesives) where good mechanical performances are not necessary. Lee et al. studied the metathesis of both dynamic bonds in PU using the same technique as shown in Fig. 10. The elastomer was formed in two stages. They began by creating a Schiff base using biomaterials such as vanillin and cystine, which is important for disulfide and imine connections [35, 37]. Before adding the PU intermediates (IPDI, isophorone diisocyanate, and poly (propylene glycol)-based thiol), the base was blended with 1,4-butanediol to confirm miscibility. Optical microscopy is used by them to track the healing process and found that using UV radiation and heat (65 °C), fractures healed completely in 120 min. In addition, the material is 97% recyclable [38].

2.3.3 Combined Covalent and Non-covalent Systems

Among the fourth generation, one of the most traverse systems is the covalent and non-covalent combination, having good efficiency of healing and mechanical properties. As a delegated combination, the generality of PU as material and disulfide bonds/hydrogen bonds can be observed as dual systems. This strategy was first applied by Rekondo et al. using the metathesis of aromatic sulfide to design PUU networking of SH. Thanks to aromatic sulfide, which at room temperature is a constant exchange, and the urea group which can form quadruple hydrogen bonds, they reached 97% healing efficiency [39]. But for high-performance applications the mechanical properties were limited. This limitation was solved by Xu et al. who developed an interesting healing system where it was assisted through sunlight, having a high tensile strength in the production of photosensitive materials with potential applications. Even though to generate disulfide oxide a pure UV was known, if the UV content is low (between 3 and 5%) in sunlight the same effect will be generated [40]. By light chromatography (HPLC) and mass spectroscopy, it's demonstrated that in small molecules the reaction of disulfides was capable of being generated by the component of UV sunlight. This exchange was assisted in this case by the creation of HB between amino groups close to disulfide main chains. With 9.5 MPa tensile strength, they found the efficiency of healing up to 96%. The production of PU with sufficient mechanical toughness has proceeded. All of these advancements in PU permitted employment in the novel application, like 3D printing for other elastomeric materials, which were previously unthinkable. Li et al. announced the creation of the first SH PUs works' digital light 3D printing (DLP), with possible applications in the production of flexible electronics and sensors. Liu et al. have expanded their research on PU below this mixture of processes of healing. Poly (vinyl alcohol)-graft-(-caprolactone) (PVA-PCL) was introduced into isocyanate terminating PU with disulfide bonding. PVA-PCL inclusion increased the production of bonds of hydrogen, which served as crosslink sites physically [41, 42]. They observed a 20 MPa increase in tensile strength and a beneficial impact of the disulfide bonds and hydrogen on the efficiency of healing, which at 90 °C reached 94%. All of the PU work demonstrates a great balance that may be achieved between healing capabilities and good performance mechanically. A general-purpose elastomer has been the second most studied system, particularly epoxidized natural rubber (ENR). Cheng et al. looked into two well-known fundamentals: the vulcanization of the main chains double bonds of cis-1,4 poly(isoprene) as well as the epoxy ring reaction. They used the standard procedure of vulcanization to build numerous sulfur linkages in the main chain, as well as disulfides of aromatic to couple the radical created during the reaction of ring-opening (Fig. 11). At the same time, the event of ring-opening produced a large number of hydroxyl groups, which formed hydrogen bindings that aided the process of SH [43]. At 120 °C this combination produced 9.30 MPa tensile and 98% healing efficiency, as compared to just 22% in an equal sample of free disulfide.

2.3.4 Hydrogen Bonds

Although the specific moment in history when humans discovered HB is still debated, it is believed that humans developed HB more than a century ago. HB is a sort of intermolecular bonding that is available in nature and is vital in activities such as the replication of DNA, protein, molecule recognition, and so on. In recent years, HB has also been used to create novel materials such as elastomers, supramolecules, and SH materials. Figure 3a shows the HB among two water molecules. The fundamental

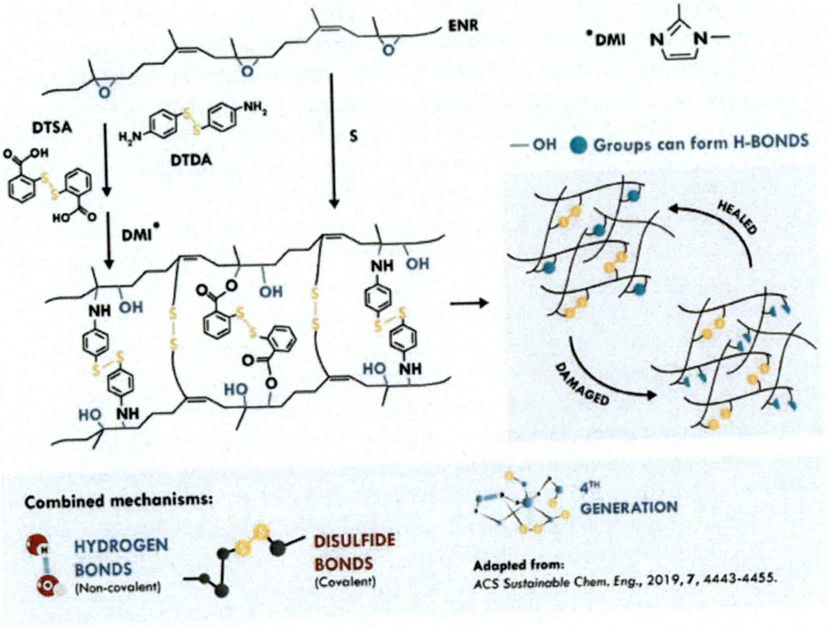

Fig. 11 Relation between hydrogen bonds and disulfide in ENR [28]

structure of HB is X–H. Y is made up of two main parts: the acceptor of a proton (with lone pair electrons in Y atom, A for short) and the proton donor (XH, D in short) [44]. A single hydrogen bond has low bond energy below 40 kJ/mol, which equates to one-tenth of binding energy for covalent (400 kJ/mol of CC bonding). As a result, we may be able to use dynamic HB to construct polymers having high SH but weak mechanical properties and tensile modulus. Even though HB can result in potent polymers with good mechanical properties comparable to hydrogen bonds, covalent polymers in such polymers are reliable, necessitating the extra energy required to fracture the prior strong interaction and afterward reconstruct an innovative HB system. Large polymers can be formed by powerful connection bonding, but their ability to self-heal is impaired. Multifunctional hydrogen with high accredited and the directed link receives a lot of attention for balancing bond strength and SH capacity [45].

Quadruple hydrogen bonds have a complicated system that must be created by two distinct molecules or by the oligomerization of a single molecule. The triple hydrogen-bonding relationship is quite strong. As shown in Fig. 12, we want to include two self-complementary triple hydrogen-bonding configurations: ADAD-DADA and AADD-DDAA [46].

The type of ADAD dimerization of the molecule results in six mutually exclusive secondary interactions resulting in the weak strength combination of the hydrogen bonds ADAD-DADA quadruple, whereas the dimer of DDAA-type molecules

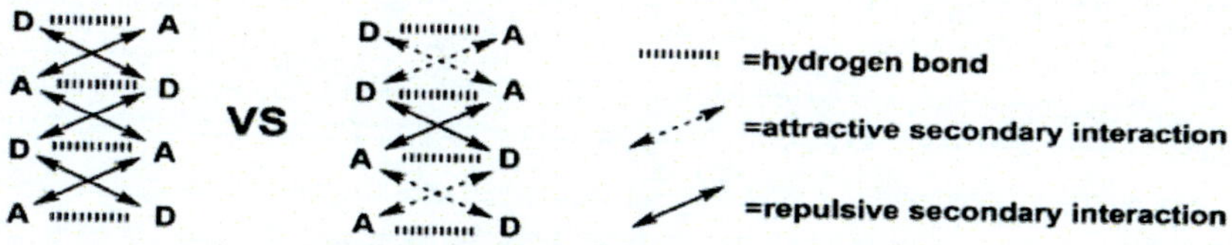

Fig. 12 Interactions and chemicals structures of DDAA and DADA dimers [46]

only has two mutually unique secondary interactions as well as four mutually hyper-attractive secondary interactions [46, 47]. As a result, the hydrogen-bonding quadruple has stronger connection in the DDAA-AADD dimer than in the dimer of ADAD-DADA.

2.3.5 Positions of H-Bonding Units in Elastomers

As previously noted, the several hydrogen-bonding dimer units can influence the characteristics of elastomers that as SH, like increased in strength mechanically, extensibility, expandability, and toughness in the resulting polymers. The density of crosslinked H-bond is determined by the location of numerous hydrogen-bonding units, which affects the properties of the ultimate material. The mechanical characteristics of various hydrogen-bonding polymers are mostly determined by the number of dissimilar hydrogen-bonding molecules and by crosslinking density [47].

2.3.6 H-Bonding Units as Chain Terminals

Because of significant constants of the heterodimer, UPy units near terminals of the chain (or ends) might be thought to increase the lengths of the main chain of the polymer, hence promoting characteristics thermally and mechanically. Furthermore, the UPy unit at the ends of the chain can provide the exceptional SH capabilities of polymers. Guan group created the first block copolymer of SH in solid form utilizing a new multiphase and biomolecule copolymer structure (poly1,2,3, Fig. 13a).

These microphase-separated molecule copolymers integrate the rigidity and toughness of thermoplastics with the dynamic and healing abilities of biomolecule materials, resulting in fast structural rigidity and elasticity recovery. Pyun et al. synthesized perfluoropolyether's (UPyPFPEs, Fig. 13b) UPy-capped that recovered the modulus of storage to its original value at 130 °C in 2 min due to the creation of crystalline structure phase separation based on UPy domains between the softer polymer matrices and the hydrogen-bonding molecules [48]. The PFPE surface modification with alkylated UPy groups, on the other hand, showed an increased time to regain the compressibility of 18 min at 110 °C, suggesting hindered crystallization.

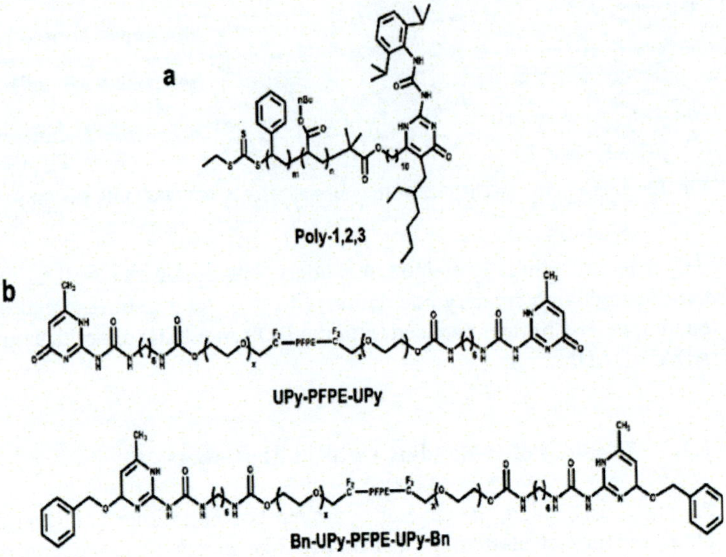

Fig. 13 Self-healing elastomers with H-bonding units as chain terminates [46]

2.3.7 Hydrogen-Bonding Units in the Main Chain

The creation of solid-like hydrogels was induced by the separation of phase domains of hydrophobic hydrogen bonding formed by UPy molecules with a PEG backbone network inside the main chain of bifunctional PEG to UPy (Fig. 14a). When the solid-like hydrogels were heated at 50 °C, they discovered SH activity. Bao and colleagues have synthesized polymeric supramolecular materials (SPMs) containing units of UPy in different concentrations (from 0 to 30 mol percent) in the main chain. Figure 14b illustrates the chemical compositions of the SPMs. The Fu group devised a unique technique based on numerous active HB to provide quick and efficient SH under room temperature or in difficult settings. Thiourea molecules were integrated into the separated microphase network of polyurea at the same time to generate multi-strength H-bonds (Fig. 14d), and hard domain crystallization was introduced with the dynamic H-bonds reversibly both in segments of soft and hard. Another elastomer based on UPy was constructed with spacers units packed loosely in a supramolecular elastomer, illustrated as shown in Fig. 14e, to resolve the mismatch of SH and strong strength at room temperature inherently. Aida report indicates strong mechanical SH materials with thiourea create an unusual zig-zag hydrogen-bonded structure (Fig. 14f).

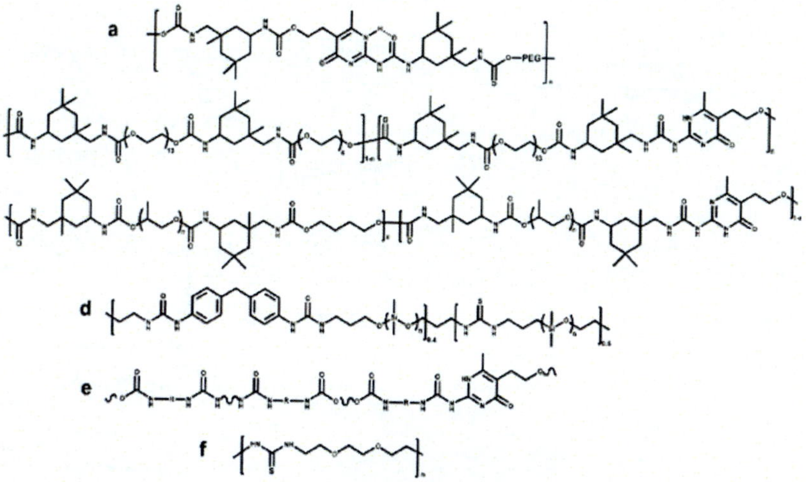

Fig. 14 SHEs with hydrogen-bonding units in the main chain (**a–f**) [46]

2.3.8 Hydrogen-Bonding Units in the Side Chain

By adding units of UPy into the side chain, Zhu et al. generated PU. They created a copolymer with a side chain of carboxyl\and then combined it with glycidol to create PU via UPy with NCO molecules. When heated, the as-synthesized PU displayed SH properties for thermal reversibility of UPy units. Lei and colleagues developed a biomass-derived elastomer for SH by combining multiple hydrogen bond formations (UPy) in the side chain with crosslinked covalent. The elastomer's dynamic nature and soft features resulted in greater extensibility and SH capacity, where the crosslink of covalent can help in the re-assembly of disrupted H-bonds (Fig. 15). As a result, the coarse elastomer had a 2600 percent break elongation and 42.76 MJ m^{-3} toughness. Furthermore, after 24 h of healing at 60 °C, the elastomer showed excellent SH ability with full scratch healing and outstanding mechanical recovery at 1900% extension and 24.1 MJ m^{-3} toughness [49].

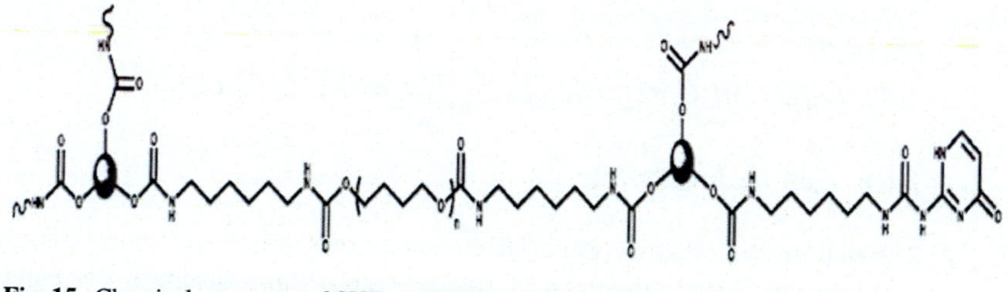

Fig. 15 Chemical structures of SHEs with H-bonding unit in the side chain [46]

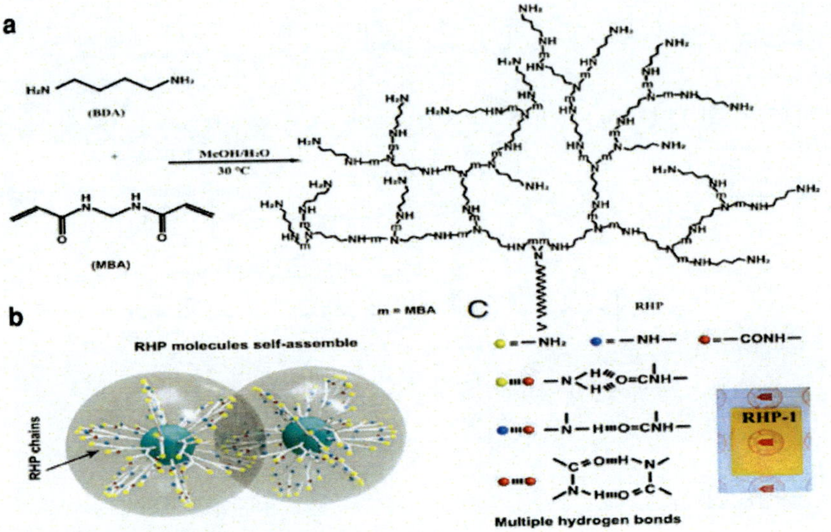

Fig. 16 Synthesis of RHPs: **a** Synthesis of the RHP, **b** Interactions between the two RHP molecules. **c** Multiple H-bonds of RHP. (Inserted) An image of (RHP-1) [46]

2.3.9 Hydrogen-Bonding Units in Branched Polymers

The Wu group created a series of random hyperbranched polymers (RHPs) with a higher density of bonds of hydrogen enabling the mechanism of SH at room temperature, as illustrated in Fig. 16. Internal molecular fragments are severely limited and have little molecular mobility in glassy hyperbranched polymers, but exterior units of branching have significant mobility in end groups [50]. As a result, the authors proposed many hydrogen-bonding groups of complimentary achieve SH by adjusting the polymer's exterior group of terminal and side chain. These polymers include many amino, amide, or other groups and are capable of forming high-density hydrogen bonds. At room temperature, 5.5 MPa tensile strength may be recovered after contact with the substance for 1 min.

2.4 Different Categories of Self-healing Elastomers

2.4.1 Self-healing PDMS

The well-known elastomer is PDMS composed of Si–O–Si molecules. Because of its exclusive structure, poly-siloxane provides compensations such as low and extreme temperature resistance, resistance to weather, ozone resistance, electrical insulation, gas permeability, hydrophobicity, bio-inertness, and non-toxicity. It's extensively used in electronics, universal health care, aircraft, textile, and other fields. Normal poly-siloxane, on the other side, does not have the same mechanical strength as polyurethane. Because poly-siloxane products are usually affected by external forces throughout the molding and usage process, scientists are striving to develop poly-siloxane elastomers having SH characteristics [51]. In the following sections, we shall explore various forms of PDMS SHEs due to several dynamic bonding mechanisms.

2.4.2 Diels–Alder Bonds

Even though a variety of SH polymeric materials based on DA connections have been developed, the large bulk of the system is PUs. SH poly-siloxane elastomers based on DA interactions are seldom described, likely due to the difficulty in identifying a suitable precursor for the production of SH PDMS. To create a successful SH PDMS, the stability of the two major pioneers, the reactivity of DA bond components (maleimide or furan) inside the predecessor, and the network of crosslinking variables such as molecular chain motility and degree of crosslinking must all be taken into account. Zhao et al. developed thermal-treatable PDMS elastomers for the first time by employing the DA reaction to crosslink polydimethylsiloxane with maleimide pendants with furan-end bifunctional siloxane [51, 52]. DA bonds could be ruptured at 140 °C and reconstructed at 80 °C. As a consequence, the healing therapy is divided into two phases: heat at 140 °C for 3 h and at 80 °C for 24 h, with a SH competence of 95%. They developed another thermally treatable poly(siloxane-urethane) elastomer (PDMS–DA–PU and PDMS/ PCL–DA–PU) to DA bonds distributed throughout the polymer chain among crosslinked locations to use self-made soft sections, isocyanate discharged polydimethylsiloxane (NCO–PDMS–NCO), or a combination of NCO–PDMS–NCO as well as polycaprolactone diol.

2.4.3 Disulfide Bonds and Hydrogen Bonds

Using a urea-containing chain-extended PDMS and multi-amino terminated hyper-branched poly-siloxane, Zhang et al. developed an innovative type of keep-going trying to expand PDMS rubber (LP2-SS-HP) with dynamic changing crosslinks due to hydrogen bonds and also chemical crosslinking induced by controlled inter-changeable aliphatic disulfide. Tg, tensile strength, elastic modulus, and hardness of customization are as strong as 96 °C, 8.6, 188.5 MPa, and 16.3 MJ m^3, correspond-ingly. Depending on tensile, elongation at break, or toughness, the cut-off films had a healing efficiency of greater than 90%. These materials have a significant advantage in that they have a high modulus and strength without compromising ductility, and they can also be mended fast following mechanical failure [53].

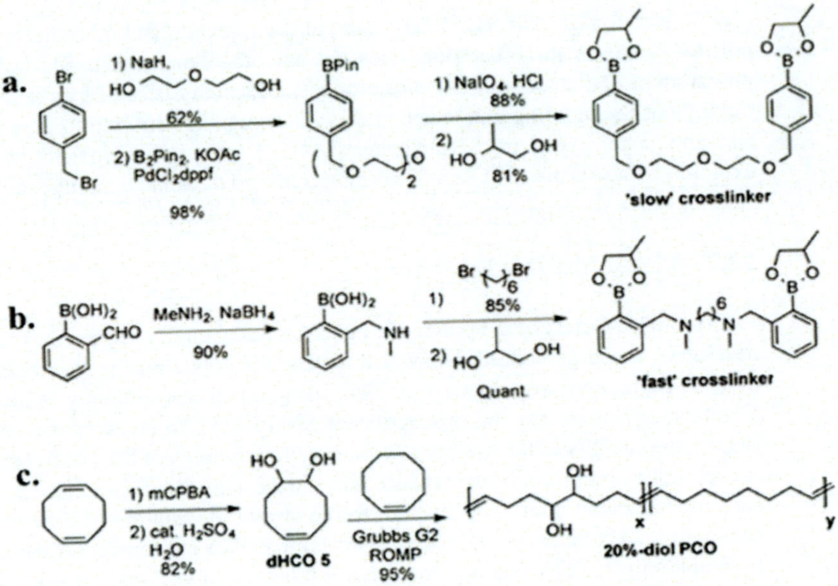

Fig. 17 Synthesis of **a** slow diphenyl-boronic ester crosslinker 3, **b** fast-di(o-amino-phenyl-boronic) ester crosslinker 4, **c** 1,2-diol-containing polycyclooctene (20% diol-PCO) [20]

2.4.4 Boronic Ester Bonds

Sumerlin et al. presented an elastomer based on thiolene reaction with outstanding SH expected to attain a dynamical boronic ester link. In their investigation, they synthesized and integrated a boronic ester diene into supporting a radically based thiolene technique. This elastomeric can be repaired at room temperature and has an efficiency of SH of 90%. However, water or a high moisture environment must be present to encourage SH action [54]. Guan et al. demonstrated how to use varying degrees of boronic ester esterification reaction to adjust the mutability and SH effectiveness of bulk polyolefin materials. They used a "slow" di phenylboronic esters crosslinker and a “fast” di (o-amino phenylboronic) ester crosslinker to crosslink 1,2-diol-containing polycyclooctene (Fig. 17a–c).

3 Applications

Contrary to conventional inorganic non-metallic and metallic materials, polymer materials own a more friendly preparation technology. Because of their functionality, multiple forms, and lightweight properties, resulting in irreversible roles in our routine life and industrial applications. Among the other polymer-based materials,

SHEs can not only increase the stability and safety of procedures. They also stimulate industrial development, prolong the service life, and have colossal commercial value. Other than this, SHEs also play an important part in expediting the effective usage of properties and the viable development of mortal society. The proclaimed SH elastomeric materials with a specific function have provoked great interest, implying the vision of effective growth.

4 Conclusions

The biological evolution, healing, and sealing wounds have been a fundamental function of the living environment, operating freely and at varied times in animals, plants, and all other living groups of things. Engineers and biologists have developed an increasing interest in these SH mechanisms as they represent a rich resource. The most common and well-known methods to gradually translate biological perceptions into technical objectives are biologically stimulated design and biomimetic access. However, just a handful of SH materials systems based on biological techniques have been developed in recent years. In contrast to the undamaged sample, these procedures can restore structural integrity and mechanical qualities as well as heal or (partially) repair damage. Most solutions presented are schematic conceptions created on a laboratory scale, but only a small number of situations can typically be realized as market items.

References

1. Ghosh, D., Sharman, R., Raghav Rao, H., Upadhyaya, S.: Self-healing systems—survey and synthesis. Decis. Support Syst. **42**(4), 2164–2185 (2007). https://doi.org/10.1016/j.dss.2006.06.011
2. Yu, K., Xin, A., Du, H., Li, Y., Wang, Q.: Additive manufacturing of self-healing elastomers. NPG Asia Mater. **11**(1) (2019). https://doi.org/10.1038/s41427-019-0109-y
3. Madsen, F.B., Yu, L., Skov, A.L.: Self-healing, high-permittivity silicone dielectric elastomer. ACS Macro Lett. **5**(11), 1196–1200 (2016). https://doi.org/10.1021/acsmacrolett.6b00662
4. Hou, J.B., et al.: Tough self-healing elastomers based on the host-guest interaction of polycyclodextrin. ACS Appl. Mater. Interfaces **11**(12), 12105–12113 (2019). https://doi.org/10.1021/acsami.9b00626
5. Kang, J., et al.: Tough and water-insensitive self-healing elastomer for robust electronic skin. Adv. Mater. **30**(13) (2018). https://doi.org/10.1002/adma.201706846
6. Gossweiler, G.R., et al.: Mechanochemically active soft robots. ACS Appl. Mater. Interfaces **7**(40), 22431–22435 (2015). https://doi.org/10.1021/acsami.5b06440
7. Zhang, Z., Qian, S., Ma, H.: Investigating mechanical properties and self-healing behavior of micro-cracked ECC with different volume of fly ash. Constr. Build. Mater. **52**, 17–23 (2014). https://doi.org/10.1016/j.conbuildmat.2013.11.001
8. Ryan, K.R., Down, M.P., Banks, C.E.: Future of additive manufacturing: Overview of 4D and 3D printed smart and advanced materials and their applications, vol. 403. Elsevier B.V. (2021)
9. Zhang, L., et al.: A highly efficient self-healing elastomer with unprecedented mechanical properties. Adv. Mater. **31**(23) (2019). https://doi.org/10.1002/adma.201901402

10. Wietor, J.L., Sijbesma, R.P.: A self-healing elastomer. Angew. Chemie Int. Ed. **47**(43), 8161–8163 (2008). https://doi.org/10.1002/anie.200803072
11. Peng, Y., et al.: Super tough and strong self-healing elastomers based on polyampholytes. J. Mater. Chem. A **6**(39), 19066–19074 (2018). https://doi.org/10.1039/c8ta06561f
12. Speck, O., Speck, T.: An overview of bioinspired and biomimetic self-repairing materials. Biomimetics **4**(1), 1–37 (2019). https://doi.org/10.3390/biomimetics4010026
13. Kim, S.M., et al.: Superior toughness and fast self-healing at room temperature engineered by transparent elastomers. Adv. Mater. **30**(1) (2018). https://doi.org/10.1002/adma.201705145
14. Guo, H., Han, Y., Zhao, W., Yang, J., Zhang, L.: Universally autonomous self-healing elastomer with high stretchability. Nat. Commun. **11**(1), 1–9 (2020). https://doi.org/10.1038/s41467-020-15949-8
15. Wu, J., Cai, L.H., Weitz, D.A.: Tough self-healing elastomers by molecular enforced integration of covalent and reversible networks. Adv. Mater. **29**(38), 1–8 (2017). https://doi.org/10.1002/adma.201702616
16. Faghihnejad, A., et al.: Adhesion and surface interactions of a self-healing polymer with multiple hydrogen-bonding groups. Adv. Func. Mater. **24**(16), 2322–2333 (2014). https://doi.org/10.1002/adfm.201303013
17. Liu, Y., et al.: A novel type of self-healing silicone elastomers with reversible cross-linked network based on the disulfide, hydrogen and metal-ligand bonds. Prog. Org. Coat. **144**(March), 105661 (2020). https://doi.org/10.1016/j.porgcoat.2020.105661
18. Dashofy, E.M., Van der Hoek, A., Taylor, R.N.: Towards architecture-based self-healing systems. Proc. ACM SIGSOFT Work. Self-Heal. Syst. 21–26 (2002). https://doi.org/10.1145/582128.582133
19. Tian, Y., et al.: Preparation and characterization of self-healing microcapsules of asphalt. Constr. Build. Mater. **263**, 120174 (2020). https://doi.org/10.1016/j.conbuildmat.2020.120174
20. Wang, Z., Lu, X., Sun, S., Yu, C., Xia, H.: Preparation, characterization and properties of intrinsic self-healing elastomers. J. Mater. Chem. B **7**(32), 4876–4926 (2019). https://doi.org/10.1039/c9tb00831d
21. Holten-Andersen, N., et al.: pH-induced metal-ligand cross-links inspired by mussel yield self-healing polymer networks with near-covalent elastic moduli. Proc. Natl. Acad. Sci. U. S. A. **108**(7), 2651–2655 (2011). https://doi.org/10.1073/pnas.1015862108
22. Gao, L., He, J., Hu, J., Wang, C.: Photoresponsive self-healing polymer composite with photoabsorbing hybrid microcapsules. ACS Appl. Mater. Interfaces **7**(45), 25546–25552 (2015). https://doi.org/10.1021/acsami.5b09121
23. Wang, H., Zhou, H., Gestos, A., Fang, J., Lin, T.: Robust, superamphiphobic fabric with multiple self-healing ability against both physical and chemical damages. ACS Appl. Mater. Interfaces **5**(20), 10221–10226 (2013). https://doi.org/10.1021/am4029679
24. Martín, R., Rekondo, A., Echeberria, J., Cabañero, G., Grande, H.J., Odriozola, I.: Room temperature self-healing power of silicone elastomers having silver nanoparticles as crosslinkers. Chem. Commun. **48**(66), 8255–8257 (2012). https://doi.org/10.1039/c2cc32030d
25. Ma, Z., Li, H., Jing, X., Liu, Y., Mi, H.Y.: Recent advancements in self-healing composite elastomers for flexible strain sensors: Materials, healing systems, and features. Sens. Actuators Phys. **329**, 112800 (2021). https://doi.org/10.1016/j.sna.2021.112800
26. Cui, M., Nguyen, N.A., Bonnesen, P.V., Uhrig, D., Keum, J.K., Naskar, A.K.: Rigid oligomer from lignin in designing of tough, self-healing elastomers. ACS Macro Lett. **7**(11), 1328–1332 (2018). https://doi.org/10.1021/acsmacrolett.8b00600
27. Lehmijoki, U., Rovenskaya, E.: Environmental Mortality and Long-Run Growth, pp. 239–258 (2010). https://doi.org/10.1007/978-3-642-02132-9_12
28. Utrera-Barrios, S., Verdejo, R., López-Manchado, M.A., Hernández Santana, M.: Evolution of self-healing elastomers, from extrinsic to combined intrinsic mechanisms: a review. Mater. Horizons **7**(11), 2882–2902 (2020). https://doi.org/10.1039/d0mh00535e
29. Dooley, C., Taylor, D.: Self-healing materials: what can nature teach us? Fatigue Fract. Eng. Mater. Struct. **40**(5), 655–669 (2017). https://doi.org/10.1111/ffe.12595

30. Kessler, M.R.: Self-healing: a new paradigm in materials design. Proc. Inst. Mech. Eng. Part G J. Aerosp. Eng. **221**(4), 479–495 (2007). https://doi.org/10.1243/09544100JAERO172
31. Jian, X., Hu, Y., Zhou, W., Xiao, L.: Self-healing polyurethane based on disulfide bond and hydrogen bond. Polym. Adv. Technol. **29**(1), 463–469 (2018). https://doi.org/10.1002/pat.4135
32. Song, T., et al\: Self-healing Materials: A Review of Recent Developments, pp. 1–19 (2021)
33. Li, X., et al.: Self-healing polyurethane elastomers based on a disulfide bond by digital light processing 3D Printing. ACS Macro Lett. **8**(11), 1511–1516 (2019). https://doi.org/10.1021/acsmacrolett.9b00766
34. Vijayasekhar, K., Acharyya, S.G., Debroy, S., Miriyala, V.P.K., Acharyya, A.: Self-healing phenomena of graphene: Potential and applications. Open Phys. **14**(1), 364–370 (2016). https://doi.org/10.1515/phys-2016-0040
35. Thakur, V.K., Kessler, M.R.: Self-healing polymer nanocomposite materials: a review. Polymer (Guildf) **69**, 369–383 (2015). https://doi.org/10.1016/j.polymer.2015.04.086
36. Tan, N.P.B., Keung, L.H., Choi, W.H., Lam, W.C., Leung, H.N.: Silica-based self-healing microcapsules for self-repair in concrete. J. Appl. Polym. Sci. **133**(12) (2016). https://doi.org/10.1002/app.43090
37. Liu, Z., et al.: Self-healing, reprocessing and 3D printing of transparent and hydrolysis-resistant silicone elastomers. Chem. Eng. J. **387**(January), 124142 (2020). https://doi.org/10.1016/j.cej.2020.124142
38. Cremaldi, J.C., Bhushan, B.: Bioinspired self-healing materials: lessons from nature. Beilstein J. Nanotechnol. **9**(1), 907–935 (2018). https://doi.org/10.3762/bjnano.9.85
39. Zhang, Z., et al.: Autonomous self-healing elastomers with unprecedented adhesion force. Adv. Funct. Mater. **31**(4), 1–10 (2021). https://doi.org/10.1002/adfm.202006298
40. Rekondo, A., Martin, R., Ruiz De Luzuriaga, A., Cabañero, G., Grande, H.J., Odriozola, I.: Catalyst-free room-temperature self-healing elastomers based on aromatic disulfide metathesis. Mater. Horizons **1**(2), 237–240 (2014). https://doi.org/10.1039/c3mh00061c
41. Hassan, B.A.R., Yusoff, Z.B.M., Othman, M.A.H., Bin, S.: A information is available at the end of the Chapter, and We are IntechOpen, the world's leading publisher of Open Access books Built by scientists, for scientists TOP 1%. Intech, p. 13 (2012). https://doi.org/10.5772/55358
42. Keller, M.W., Whitet, S.R., Sottos, N.R.: An elastomeric self-healing material. In: Proceedings of 2006 SEM Annual Conference and Exposition on Experimental and Applied Mechanics, vol. 1, no. January, pp. 379–382 (2006)
43. Arslan, M., Kiskan, B., Yagci, Y.: Benzoxazine-based thermosets with autonomous self-healing ability. Macromolecules **48**(5), 1329–1334 (2015). https://doi.org/10.1021/ma5025126
44. Zeng, C., Seino, H., Ren, J., Hatanaka, K., Yoshie, N.: Self-healing bio-based furan polymers cross-linked with various bis-maleimides. Polymer (Guildf) **54**(20), 5351–5357 (2013). https://doi.org/10.1016/j.polymer.2013.07.059
45. Ahn, T.H., Kishi, T.: Crack self-healing behavior of cementitious composites incorporating various mineral admixtures. J. Adv. Concr. Technol. **8**(2), 171–186 (2010). https://doi.org/10.3151/jact.8.171
46. Xie, Z., Hu, B.L., Li, R.W., Zhang, Q.: Hydrogen bonding in self-healing elastomers. ACS Omega **6**(14), 9319–9333 (2021). https://doi.org/10.1021/acsomega.1c00462
47. Garcia, S.J.: Effect of polymer architecture on the intrinsic self-healing character of polymers. Eur. Polym. J. **53**(1), 118–125 (2014). https://doi.org/10.1016/j.eurpolymj.2014.01.026
48. Lai, Y., Kuang, X., Zhu, P., Huang, M., Dong, X., Wang, D.: Colorless, transparent, robust, and fast scratch-self-healing elastomers via a phase-locked dynamic bonds design. Adv. Mater. **30**(38), 1–8 (2018). https://doi.org/10.1002/adma.201802556
49. Yang, R., et al.: Constructing electrically and mechanically self-healing elastomers by hydrogen bonded intermolecular network. Langmuir **36**(12), 3029–3037 (2020). https://doi.org/10.1021/acs.langmuir.0c00221
50. Zhang, L., Bailey, J.B., Subramanian, R.H., Tezcan, F.A.: Hyperexpandable, self-healing macromolecular crystals with integrated polymer networks. Nature **557**(7703), 86–91 (2018). https://doi.org/10.1038/s41586-018-0057-7

51. Qian, S.Z., Zhou, J., Schlangen, E.: Influence of curing condition and precracking time on the self-healing behavior of Engineered Cementitious Composites. Cem. Concr. Compos. **32**(9), 686–693 (2010). https://doi.org/10.1016/j.cemconcomp.2010.07.015
52. Moynihan, D.P.: Family and nation revisited. Soc. Thought **16**(2), 49–60 (1990). https://doi.org/10.1080/15426432.1990.10383700
53. Psaier, H., Dustdar, S.: A survey on self-healing systems: approaches and systems. Comput. (Vienna/New York) **91**(1), 43–73 (2011). https://doi.org/10.1007/s00607-010-0107-y
54. Azcune, I., Odriozola, I.: Aromatic disulfide crosslinks in polymer systems: self-healing, reprocessability, recyclability and more. Eur. Polym. J. **84**, 147–160 (2016). https://doi.org/10.1016/j.eurpolymj.2016.09.023

Synthetic-Origin Biodegradable Polymers

Muzammil Mehmood, Ahsan Ahmad, and Muhammad Tahir Khan

Abstract The utilization of non-biodegradable disposable plastic has increased exponentially due to its lightweight and user-friendliness. Nonetheless, the non-biodegradable plastic waste produced by disposable plastics increases the environmental carbon footprint. The utilization of biodegradable plastics is one solution to mitigate the environmental pollution problem. Currently, only 1% of 300 million tons of plastic produced annually is biodegradable. Biodegradable plastics are obtained from both natural and synthetic resources. Due to their lower strength, natural-origin biodegradable plastics are only used in a few applications. For a large number of applications, the mechanical strength and other properties of synthetic-origin biodegradable plastics can be engineered. This chapter covers the biodegradable plastics of synthetic-origin including aliphatic polyesters, aromatic copolymers, vinyl polymers, and biodegradable polyurethanes. Biodegradation (both biotic and abiotic) in synthetic-origin biodegradable polymers has been discussed followed by the classification and degradation tendencies in synthetic-origin biodegradable plastics.

Keywords Biodegradable polymers · Synthetic origin · Aliphatic polyesters · Aromatic co-polyesters

1 Why the Planet Needs Biodegradable Plastics?

Modern lifestyle is highly dependent on the use of plastic materials due to their versatility and cost-effectiveness. A majority of such plastics are polyolefins derived from petroleum-based resources. Since these resources are finite in quantity, their sustained supply cannot be assured. Moreover, the end-of-life plastic is deposited into the environment as persistent pollutants posing a serious threat to our ecosystem.

Global plastic production was 367 million metric tons in the year 2020 and is expected to rise to over 445 million metric tons by the year 2025 [1]. Global plastics

M. Mehmood (✉) · A. Ahmad · M. T. Khan
Department of Materials, National Textile University, Faisalabad, Pakistan
e-mail: muzammil@ntu.edu.pk

K. Shaker and A. Hafeez (eds.), *Advanced Functional Polymers*, Engineering Materials,
https://doi.org/10.1007/978-981-99-0787-8_7

consumption is always higher than that produced in a year because a certain quantity is also recycled from the past years. Out of all the end-of-life plastic produced globally, 40% is dumped into landfills, 25% is combusted, 12% is subjected to mechanical recycling, 1% is utilized in monomer recovery, and 22% remains unattended [2].

Landfilling is a cheaper opportunity for managing plastic waste, but it may result in a bunch of undesirable effects. The undesirable effects include groundwater and air pollution, soil contamination, and disturbance in the natural environment. Landfilling broadly falls into two major categories, namely, (i) sanitary or managed landfills and (ii) non-sanitary or unmanaged dumps. In sanitary landfills, bottom liners and top covers are essentially required to minimize the penetration of leachates into the soil. Bottom liners and top covers also minimize the unpleasant odor and dust. In non-sanitary or unmanaged dumps, no bottom liners and top covers are required. In unmanaged dumps, the plastic waste is dumped in holes extending below the groundwater level. Pre-treatment of plastic waste is necessary for both managed and unmanaged landfilling. If the plastic waste is landfilled without pre-treatment or segregation, highly toxic leachates are produced. These leachates eventually reach humans through the food and water chain. Additionally, landfilling of untreated plastic waste results in longer (up to several hundred years) land-reclamation times [3].

Incineration focuses on the reduction of the quantity of plastic waste since it is a material destruction technique rather than a material recovery technique. Incineration results in the emission of a variety of hazardous and toxic organic pollutants. The organic pollutants include brominated and polychlorinated compounds, dioxins, furans, soot, and ash. Persistent exposure to the aforementioned pollutants is severely damaging to humans, animals, and vegetation. The brominated compounds enter humans and animals via the respiratory tract and cause carcinogenesis and mutagenesis. Likewise, the dioxins enter humans and animals via the water and food chain ultimately causing damage to reproductive and respiratory systems [4].

Recycling plastic waste can remarkably reduce environmental pollution and consumption of crude oil. Collection, sorting, and pre-treatment are essentially required for the effective recycling of plastic waste. Plastic waste recycling can be advantageous only after the implementation of an effective waste collection strategy. To avoid the generation of undesirable byproducts, plastic waste needs to be segregated before recycling [5].

All of the unattended end-user plastic waste is accumulated in the ecosystem for being non-biodegradable. A significant part of unattended plastic waste strews about on land and leaks into the ocean and rivers. An additional increase of 15% in plastic waste quantity each year brings further pollutes the ecosystem. With an estimated 8 million tons of plastic waste discarded annually into the oceans, there may be more plastic in the ocean than fish by 2050.

All the aforementioned shortcomings associated with conventional plastics require the world to pay more attention to the production and utilization of bioplastics. Bio-based plastics can assist the polymer industry to improve its sustainability and reduce crude oil dependency. Due to the environmental compatibility, bioplastics (both natural and synthetic) are the solution to environmental challenges [6].

2 Introduction to Biodegradability

Biodegradability or biodegradation is the breakdown of polymer molecules into salts, gases, and biomass. Biodegradation occurs under the influence of biotic and abiotic factors. The factors include algae, fungi, bacteria, moisture, air, sunlight, ultraviolet radiation, temperature, and soil. Some biodegradable polymers require a combination of these factors to degrade. A biodegradable polymer usually undergoes (i) erosion or fragmentation, (ii) depolymerization, and (iii) mineralization to completely degrade into non-toxic substances.

Biodegradable polymers originated in the early 1980s. Biodegradable polymers can cleave or erode themselves after a specific period under specific conditions. Biodegradability in polymers is attributed to the presence of hydrolyzable and oxidizable linkages in the backbone. While natural biodegradable polymers (NBPs) inherently possess this capability, synthetic biodegradable polymers (SBPs) are purposefully engineered for this capability. Both renewable and non-renewable resources can be utilized for the production of biodegradable polymers [7].

Biodegradation in polymers can occur aerobically or anaerobically. Aerobic biodegradation (in the presence of oxygen) produces carbon dioxide along with water, biomass, and residue. Anaerobic biodegradation yields methane gas instead of carbon dioxide, along with water, biomass, and residue. For a polymer to be completely biodegradable, aerobic and anaerobic biodegradation must be followed by mineralization. The process of mineralization refers to the complete transformation of produced biomass into gases, water, minerals, salts, and residue [8].

The extent or degree of biodegradability in polymers can be measured by various approaches. If the process of mineralization is complete, the amount of biomass produced can be used as a measure of biodegradability. For oxo-biodegradable polymers, the rate of oxygen uptake gives the measure of biodegradability. Aerobically biodegradable polymers yield carbon dioxide upon biodegradation and the quantity of carbon dioxide produced in a certain amount of time may be used as a measure of biodegradability. Similarly, the products of hydrolysis or other chemical reactions occurring during biodegradation may be used to assess the degree of biodegradability.

Fourier transform infrared spectroscopy (FTIR), ultraviolet spectroscopy (UV), or nuclear magnetic resonance spectroscopy (NMR) are commonly employed for measuring biodegradability. During biodegradation, the mechanical and physical properties of polymers are reduced and the quantitative value of the reduction in properties represents the extent of biodegradability. For measuring the reduction in mechanical properties, a universal test frame (UTM) or dynamic mechanical analyzer (DMA) is most frequently used. The reduction in physical properties is measured using an X-ray diffraction analyzer (XRD), Thermo gravimetric analyzer (TGA), gel permeation chromatography (GPC), and differential scanning calorimeter (DSC) [9].

Keeping the physical factors aside, biodegradability can broadly be classified into biotic and abiotic degradability. Biotic degradation is always predominant over abiotic degradation. Although some physical and abiotic factors are essentially involved in biotic biodegradation.

2.1 *Biotic Degradation*

Biodegradation that takes place in the presence of oxygen by the microorganism is termed biotic degradation. The microorganisms recognize the subject polymer molecule as feed and yield non-toxic products of biodegradation along with the energy. The liberated energy is then utilized by the microorganisms for their life functioning. The process of biodegradability is accomplished by the microorganisms in three stages. The first stage corresponds to the reduction in molecular weight while the second stage is attributed to the reduction in crystallinity of the polymer. The mechanical properties of the subject polymer are reduced in the third and final stage of biodegradation.

Biotic degradation can be further classified as physically, chemically, and enzymatically induced biotic degradation.

2.2 *Abiotic Degradation*

Abiotic degradation takes place in the absence of oxygen by abiotic factors like mechanical stresses, heat, sunlight, soil, and chemicals. Abiotic factors usually commence the process of biodegradation by dwindling the polymeric structure [10]. Abiotic degradation is categorized into the following sub-classes.

2.2.1 Mechanical Stress-Induced Abiotic Degradation

Abiotic degradation at the molecular level is initiated by tensile, compressive, and shear forces in polymers. The forces may arise from a variety of factors including but not limited to air and water-induced turbulences, aging, mechanical loading, and soil-burial pressures. The stress-induced abiotic degradation is only evident at the microscopic level. Only the macroscopic level stress-induced abiotic degradation is observable by the naked eye. The microscopic level stress-induced changes, when aided by other factors, usually result in defragmentation or rupture [11].

2.2.2 Temperature-Induced Abiotic Degradation

Thermally-induced abiotic degradation takes place at temperatures above the melting points in polymers. A solid-to-liquid phase transformation is always involved in this type of abiotic degradation. Since the environmental temperature in most parts of the world is below the melting temperature of thermoplastic polymers, biodegradation is not initiated at ambient conditions. Some polymeric composites, however, possess a melting point near the ambient temperature. The macromolecular structure and the crystallinity are altered when the polymers are heated near the glass transition temperature. The aspect is typically evident in semi-crystalline polymeric materials containing both crystalline and amorphous regions in their structure. Such materials when heated beyond their glass transition temperature (T_g) become more vulnerable to biotic or abiotic degradation. The addition of transition metal ions in polymers serves as pro-oxidants. Thermo-oxidative abiotic degradation is depicted by the polymers containing such pro-oxidants. For instance, the thermo-oxidative degradation in pro-oxidant-containing polyethylene (PE) films is initiated and controlled by varying the temperature [12].

2.2.3 Sunlight and UV Induced Abiotic Degradation

In some polymers, abiotic degradation is triggered by exposure to sunlight or ultraviolet (UV) radiation. When the polymers containing UV-sensitive functional groups are exposed to ultraviolet radiation, a Photolysis reaction is initiated. Likewise, the presence of photoinitiators in the polymer results in the initiation of an oxidation reaction. The incorporation of carbonyl groups in the polymer backbone usually imparts UV-sensitivity to the polymer. The copolymer of ethylene and carbon mono oxide depicts UV-sensitivity due to the presence of the carbonyl group in the backbone. Despite being photo-degradable, the carbonyl-containing polymers are more expensive than their non-carbonyl-containing competitors. A cost-effective method to inculcate photo biodegradability in polymers is the addition of non-toxic transition metal compounds. Pro-oxidant nature is also imparted to polymers by the addition of transition metal compounds. Polymers with pro-oxidant nature are unstable at processing conditions due to the initiation of thermally induced oxidation during processing. UV stabilizers and antioxidants are added to polymers to prevent undesirable thermally induced oxidation. The addition of antioxidants and UV stabilizers also provides an opportunity to engineer the lifespan of the polymer from a few weeks to several years. The lifespan of secondary food packaging films is controlled by this phenomenon [13].

2.2.4 Chemically Induced Abiotic Degradation

Chemically induced abiotic degradation is predominantly the result of oxidation or hydrolysis. The chemical pollutants, agrochemicals, molecular oxygen, and traces of ozone present in the atmosphere result in the oxidative degradation of polymers. The molecular oxygen present in the atmosphere attacks the covalent bonds in polymers and produces free radicals. Chain scissions and crosslinking are initiated by the generated free radicals [10]. Oxidative degradation in the polymer can be induced either catalytically or non-catalytically. In a polymer, it is possible to carry out abiotic oxidative degradation at ambient or elevated temperatures (both above and below the glass transition temperature T_g). The rate of oxidative degradation in polymers can be enhanced by coupling different modes of oxidation. For this purpose, chemical oxidation is usually coupled with ultrasonic-assisted oxidation, UV/sunlight photo-oxidation, thermal oxidation, or nuclear radiation-assisted oxidation. Some polymers exhibit an induction period upon exposure to oxygen during which oxidative degradation is absent. No structural change or oxygen absorption occurs during the induction period. The exposure of such polymers to oxygen during the induction period usually results in the formation of hydroperoxides responsible for rapid auto-oxidation at later stages. The oxidative degradation starts immediately in the polymers containing metallic salts due to the absence of an induction period. The oxidation of polymers is usually accomplished in three steps namely the initiation, the propagation, and the termination step.

The initiation step is characterized by the formation of free radicals either by physical (UV radiation, temperature, or ultrasonics) or chemical (peroxides, hydroperoxides, inhibitors, solvent, catalyst, and atomic state) factors. The propagation step is characterized by a variety of transformations including the formation of multiple functional groups and radicals. Firstly, the macro-sized radicals formed during the initiation step are further oxidized to peroxy radicals. The conversion of macro-sized free radicals to peroxy radicals is inversely related to the stability of free radicals. The peroxy radical formed further reacts with the polymer substrate to preferentially eliminate the secondary and tertiary hydrogen from the polymer. When the eliminated hydrogen atoms interact with the free radicals, hydroperoxides are formed. The formation of hydroperoxides is usually favored by low temperatures for being exothermic. Further decomposition of hydroperoxides then yields the alkoxy radicals and the hydroxyl groups. These alkoxy radicals give rise to polymer chain scission thereby directly contributing to fragmentation. The free radicals obtained during the initiation and propagation step combine to form inactive products thereby resulting in the termination of the reaction [14].

Hydrolytic degradation refers to the cleavage of water-sensitive functional groups in the polymer backbone for a partial or complete degradation of the polymer. Hydrolytic degradation is a combination of a physical and chemical phenomenon, with the physical phenomenon referring to the absorption of water by polymer molecules while the chemical phenomenon refers to the cleavage of hydrolyzable functional groups. Hydrolytic degradation is referred to as "bulk degradation" if the physical water absorption by the polymer is quicker than the chemical bond breaking. Most of the hydrolytically degradable polyesters, polyglycolic acids, and polylactic acids fall under this class. Contrastingly, if the polymer linearly unhooks its mass due to the hydrolytic degradation occurring at the polymer-aqueous interface only, it is referred to as "surface erosion degradation." This phenomenon is mostly predominant in polyanhydrides [15]. Some biodegradable polymers may undesirably undergo hydrolytic degradation during their processing due to high humidity or temperature. The phenomenon can occur even at moderate processing temperatures leading to difficulties in their processing. This undesirable effect can be countered by incorporating modifying chemicals like carbodiimides and epoxides in the polymers [16].

3 Classification of Synthetic-Origin Polymers

The synthetic-origin polymers are categorized mainly into two types, i.e., biodegradable, and non-biodegradable as per their synthesis and processing method, chemical composition, economic importance, and end applications. In this chapter, synthetic-origin biodegradable polymers will be discussed in detail owing to their importance in recent environmental pollution issues and concerns. Classification of biodegradable and non-biodegradable common synthetic-origin polymers is given in Fig. 1.

3.1 Synthetic-Origin Biodegradable Polymers

Synthetic-origin polymers with hydrolyzable backbones are prone to biodegradation under specific conditions. The class of polymers having these types of chain backbones includes aliphatic polyesters, aromatic copolyesters, some vinyl polymers (e.g., PVA and PVAc), and polyurethanes (Fig. 1).

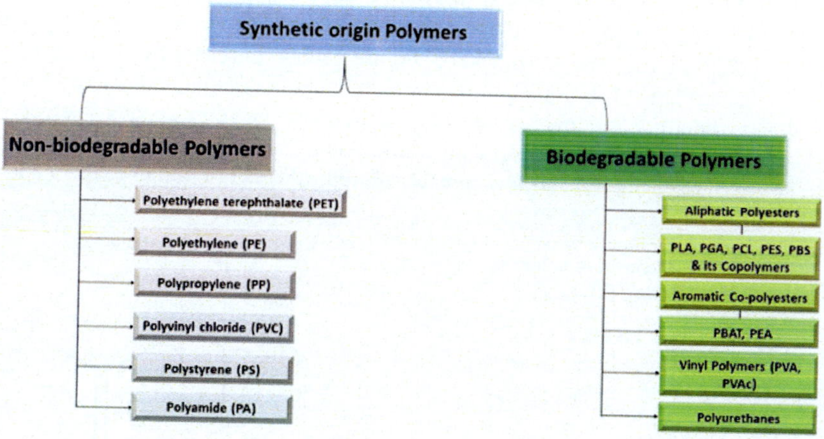

Fig. 1 Classification of common biodegradable and non-biodegradable synthetic-origin polymers

3.1.1 Aliphatic Polyesters

They are the most extensively studied and in-use class of synthetic-origin biodegradable polymers, from the polyesters family because of their versatility and diversity. In the process of their development, various monomers from large varieties and different polymerization routes can be in consideration. If we are discussing the polymerization routes, then for high molecular weight polymers like polyesters, ring-opening polymerization is chosen over polycondensation [17]. Six-membered lactones are used in ring-opening polymerization (ROP) for the preparation of most biodegradable polyesters. Being the only high molecular weight biodegradable polymers class, aliphatic polyesters are broadly explored, and their hydrolyzable ester bonds give biodegradability [18].

They can be categorized into two different types related to the bonding mechanism of their integrant monomers. Polyhydroxyalkanoates are the first type synthesized by the polymerization of hydroxy acids (HO-R-COOH). Some of the examples are poly (glycolic acid) (PGA) and poly(lactic acid) (PLA). Polymers of the second type are poly (alkene dicarboxylate)s and prepared by the polycondensation reaction of dicarboxylic acids and diols. Examples are poly(ethylene succinate) (PES) and poly(butylene succinate) (PBS) [19].

3.1.2 Polylactic Acid (PLA)

Polylactic acid is a linear aliphatic, thermoplastic polyester synthesized normally using cyclic dilactone by ring-opening polymerization capable of renewability along with high biocompatibility, UV stability, and strong rigidity [20]. Various structural parameters of PLA like composition (stereo-regularity content) and molecular weight (Mw) may decide thermal properties. It is a semi-crystalline or near amorphous polymer having a glass transition temperature (T_g) of 55 °C and melting temperature (T_m) ranging from 180 to 190 °C. Figure 2 shows the chemical structure of Polylactic acid (PLA) [21].

Fragility and low resistance to temperature are some apparent defects that PLA suffers usually. For addressing some other disadvantages are intrinsic brittleness, water sensitivity, and low impact strength, the use of suitable fibers and fillers in

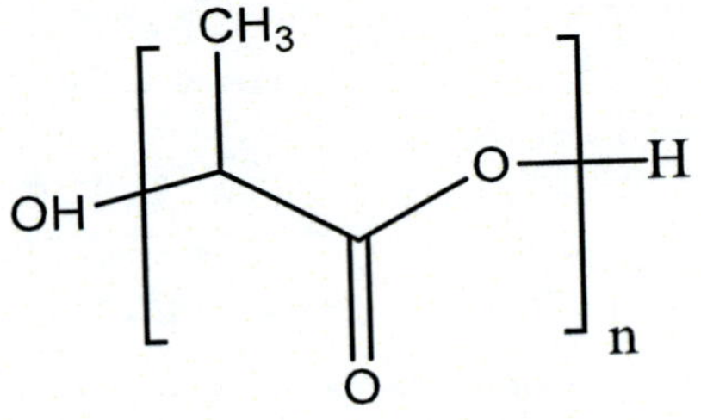

Fig. 2 Chemical structure of polylactic acid (PLA)

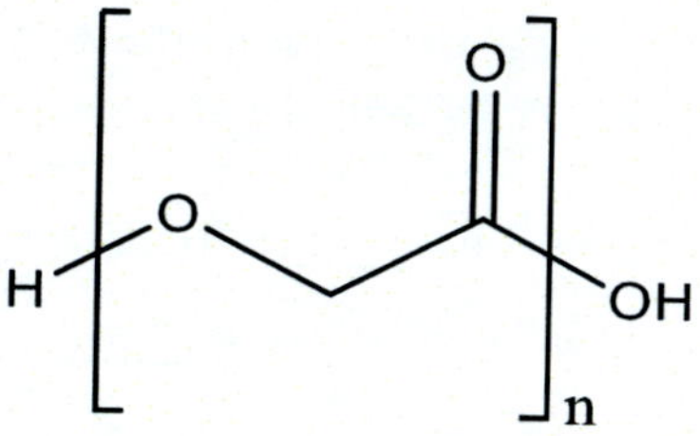

Fig. 3 Chemical structure of Polyglycolic acid (PGA)

the processing of PLA is an efficient and convenient way of enhancing the overall properties [22].

In recent years consumption of PLA has increased because of advanced manufacturing technologies results enhanced production rates due to which a gradual decrease in price for PLA was observed [23]. It can be used for the production of suitable biomaterials for several different applications by using its excellent mechanical properties and remarkable barrier capability.

3.1.3 Polyglycolic Acid (PGA)

PGA is a simple linear aliphatic polyester, usually synthesized by the ring-opening polymerization of glycolide and cyclic lactone and has a chemical structure (Fig. 3) identical to polylactic acid (PLA) but without having a methyl side group along the main chain, that brings a high degree of crystallinity around (45–55%) due to the tight packing of polymer chains, results in high melting temperature (thermal stability) about (Tm = 220–230 °C), extraordinarily high gas barrier properties (more than 3 times higher to EVOH), along with high stiffness (7 GPa) and mechanical strength (115 MPa) [24].

On the other hand, PGA normally shows a similar rapid biodegradation trend along with a 100% composability ratio as that of cellulose. Till now, PGA has been mostly used in its copolymer form, like poly(lactic-co-glycolic acid) (PLGA). Its distinctive properties have often been neglected and so far, not been fully investigated yet. The reason behind this is intrinsic properties associated with it such as rapid degradation profile, brittle behavior, high hydrophilicity, and insolubility issue in many organic solvents that have obstructed its applications on practical grounds [25].

Besides all these limitations PGA finds applications in many active fields today such as in drug delivery [26] and tissue engineering [27] because of its human's friendly nature, and functional use like in sensors, coating, packaging, etc.

3.1.4 Polycaprolactone (PCL)

PCL is usually produced by the ring-opening polymerization of comparatively low-cost cyclic monomer known as ε-caprolactone and in the existence of a catalyst called

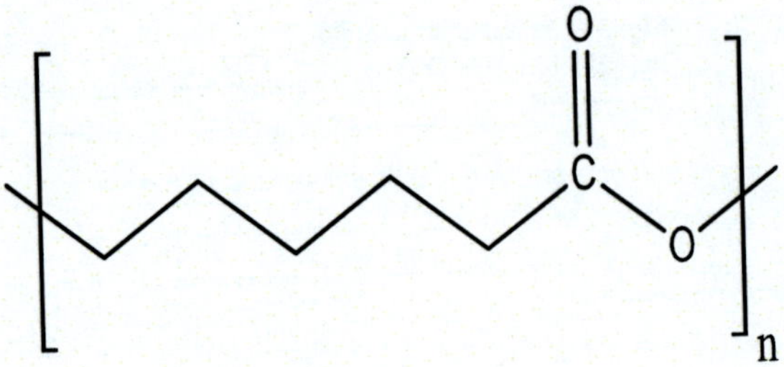

Fig. 4 Chemical structure of polycaprolactone (PCL)

tin octoate (Tin(II) 2-ethyl hexanoate) [28]. The chemical structure of PCL can be seen in Fig. 4.

PCL is an extremely tough aliphatic linear polyester and semi-crystalline in nature having adequate biocompatibility and biodegradability. It can be soluble in a wide range of inorganic and organic solvents. Being a temperature-sensitive biodegradable polyester, PCL has a low glass transition temperature (Tg) of about −60 °C and melting temperature (Tc) around 60–65 °C. Because of its low Tg, PCL is used as a soft block or as a compatibilizer in different polyurethane formulations [29].

At room temperature, it acts as a semi-rigid material and has lower values of tensile strength (i.e., 23 MPa) and relatively higher elongation to break values (more than 700%), if we are talking about the modulus of PCL it usually lies in between high-density polyethylene (HDPE) and low-density polyethylene (LDPE).

Crystal dominations remain intact during the degradation process of PCL because it usually takes place in amorphous constituent domains. It can easily be degradable under the action of microorganisms like enzymes and fungi. To improve the degradation pace, several copolymers of PCL can be made with lactide or glycolide [30].

3.1.5 Polyethylene Succinate (PES)

PES is usually synthesized either by ring-opening polymerization (ROP) of succinic anhydride with ethylene oxide or polycondensation reaction of succinic acid and ethylene glycol [31]. The chemical structure of polyethylene succinate (PES) can be seen in Fig. 5.

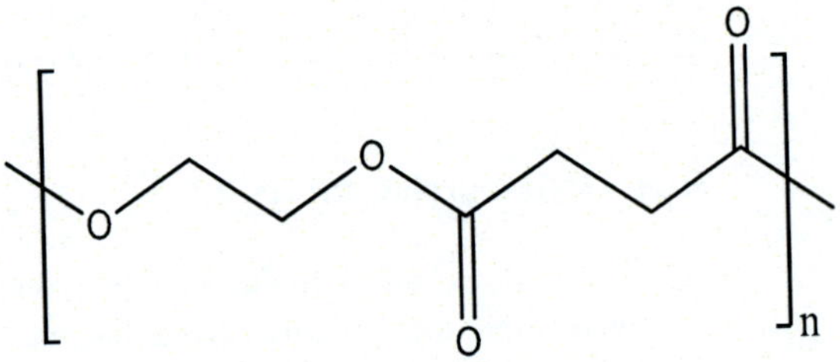

Fig. 5 Chemical structure of polyethylene succinate (PES)

It has a melting temperature (Tm) in the range of 103–106 °C, with many good mechanical properties, specifically elongation. For film making and utility, it can be considered in application because of its high-level gas barrier properties along with excellent biodegradability.

3.1.6 Polybutylene Succinate (PBS) and Its Copolymer

A white crystalline aliphatic thermoplastic polyester is generally synthesized by the polycondensation reaction of succinic acid with 1,4-butanediol. PBS has glass low transition temperature (Tg) of about −45 to −10 °C and melting temperature (Tc) around 90–120 °C. PBS is chemosynthetic polyester in nature and its melting, as well as crystallization behavior, is well reported in the literature [32]. The chemical structure of Polybutylene succinate (PBS) can be visible in Fig. 6.

Mechanical properties like tensile strength are 32 MPa and elongation at break is about 330%. These are comparable with polyethylene and polypropylene which are widely used polymers. Other interesting properties such as chemical and thermal resistance, melt processability, and most importantly biodegradability [33].

Because of its excellent processability, it can be used in the polymer processing field easily in the temperature range of 160–200 °C. Applications in all related products including containers, mulch film, cutlery, bags, packaging film, and flushable hygiene products. Due to the high crystallization rate and crystallinity, it has a comparatively low biodegradation rate. To overcome this issue several approaches have been adopted in this regard, such as the formation of composites, copolymerization, or physical blending [34].

Polybutylene succinate adipate (PBSA):

Low biodegradation rates of polybutylene succinate (PBS) urged scientists to work on copolymerization. PBSA is a copolymer of PBS synthesized by the addition reaction of aliphatic dicarboxylic acids with glycols. Figure 7 shows the chemical structure of polybutylene succinate adipate (PBSA). The presence of succinic acid (prepared by the fermentation of sugar obtained from sugarcane) confirms its biomaterial class [35].

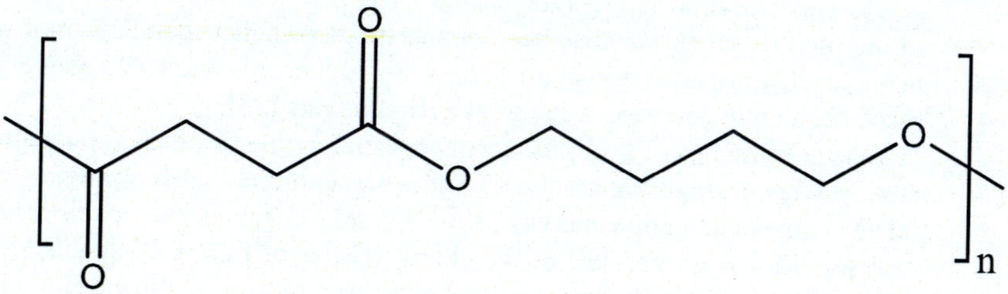

Fig. 6 Chemical structure of polybutylene succinate (PBS)

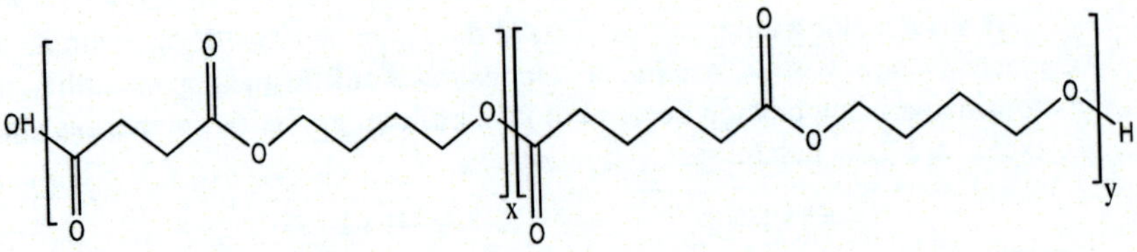

Fig. 7 Chemical structure of polybutylene succinate adipate (PBSA), (here x = 4, y = 2, 4)

It can be used in different applications like injection molded cutlery, sheet extrusions, blow-molded containers, foam cushions, monofilaments, multifilaments, and films. For composting purposes, PBSA film is suitable especially for kitchen waste because it has properties closer to linear low-density polyethylene (LLDPE) and has a relatively high biodegradability rate [36].

3.2 *Aromatic/Aliphatic Co-polyesters*

Using aliphatic monomers of different sizes, a large number of polyesters or copolyesters have been synthesized in recent years. However, their mechanical properties are lower in range than those of conventional non-biodegradable polymers. They typically show sensitivity toward microbial or enzymatic attack and hydrolytic degradation. To overcome this issue, a combination of aliphatic–aromatic copolyesters was introduced. Aromatic and aliphatic monomers are used in a mixture and are usually based on terephthalic acid [37].

3.2.1 Polybutylene Adipate Terephthalate (PBAT)

PBAT is one of the most regularly studied and used copolyesters. A mixture of terephthalic acid and adipic acid along with 1,4-butanediol is used in the synthesis of PBAT and done by a polycondensation reaction. The chemical structure of polybutylene adipate terephthalate (PBAT) is visible in Fig. 8.

The concentration of terephthalic acid can play a vital role in the overall properties of the final polymer. For instance, a concentration higher than 35% mol will give better mechanical and thermal properties. But higher concentrations above 55% mol results in a rapid decrease in the biodegradation rate [38].

Like polyethylene, PBAT has lots of appealing properties such as puncture toughness and modest high impact. Low stiffness and normal tensile strength along with relatively high elongation at break (30–40%) are some main characteristics of PBAT reported. Moreover, because of the characteristics of fully biodegradable material (compostable), it can be managed and processed on typical blown film equipment usually employed for polyethylene. PBAT can work as a toughening agent in the processing of poly(lactic acid) PLA [39].

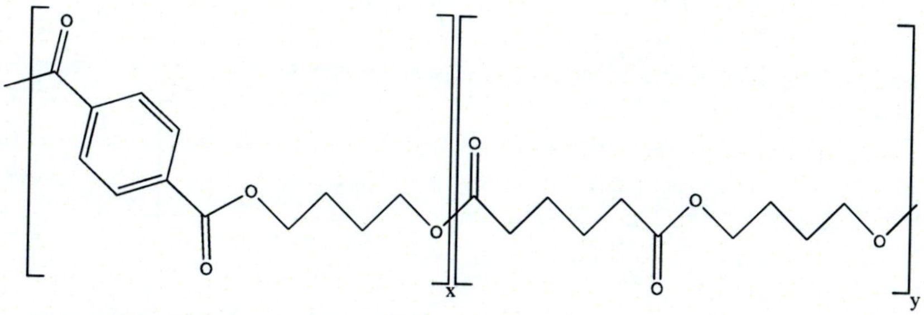

Fig. 8 Chemical structure of polybutylene adipate terephthalate (PBAT)

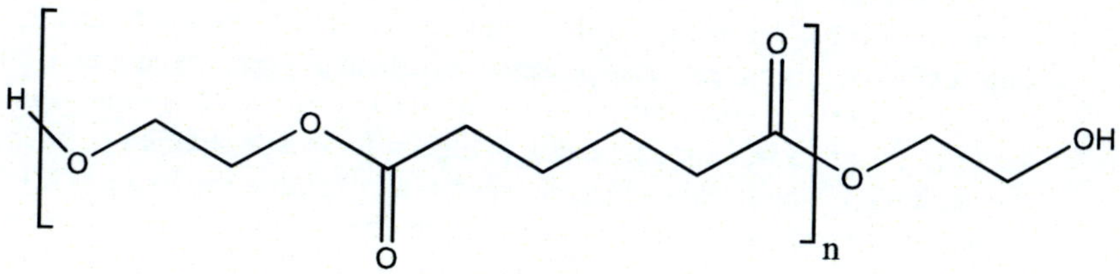

Fig. 9 Chemical structure of polyethylene adipate (PEA)

3.2.2 Polyethylene Adipate (PEA)

It is from the class of aliphatic copolyesters. Mostly synthesized using adipic acid and ethylene glycol as base units by choosing polycondensation reaction. The chemical structure of polyethylene adipate (PEA) is presented in Fig. 9.

Because of its biodegradable nature, through various mechanisms and as compared to other common polymers PEA is relatively inexpensive. The reason behind biodegradability is its lower molecular weight compared to many other polymers [40].

3.3 Vinyl Polymers

3.3.1 Polyvinyl Alcohol (PVA)

It is the most quickly biodegradable polymer from the vinyl polymers family and is known as "Bioplastic." Wastewater-activated sludge easily degrades it. In contrast to many vinyl polymers, polymerization of PVA has not been done usually from its subsequent monomer. Wholly, acetaldehyde and tautomeric forms of vinyl alcohol and monomer exist. Chemical structural depiction of polyvinyl alcohol (PVA) available in Fig. 10.

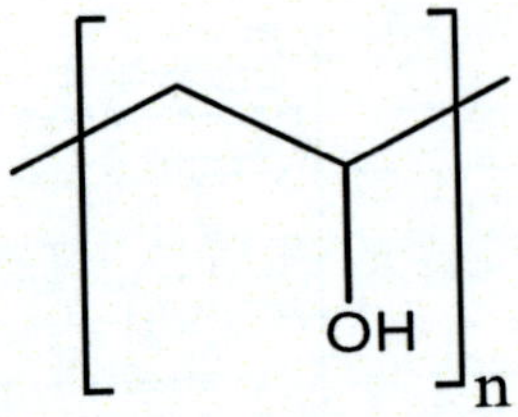

Fig. 10 Chemical structure of polyvinyl alcohol (PVA)

PVA has a molecular weight ranging between 26,300 and 30,000 with a melting point of around 180–190 °C precisely and values 86.5–89% degree of hydrolysis [41]. The film-forming, emulsifying, and adhesion capabilities of PVA are exceptional. It also has oil, grease, and solvent resistance. In addition to its high tensile and flexibility, this material also possesses excellent oxygen and odor barrier characteristics. However, these attributes are influenced by humidity, in other words, more water is absorbed when the humidity is higher. Its tensile strength is reduced with the addition of water, while elongation and tear strength are enhanced [42].

As far as synthetic water-soluble polymers in concern, PVA is the most widely available and largely produced polymer on the planet. PVA's biodegradability in the environment is one of the most notable features of its properties set. The two-step biodegradation reaction involving the oxidation of the hydroxyl group and subsequent hydrolysis is the most widely accepted mechanism. Because of its excellent biodegradability and mechanical characteristics, PVA has received a great deal of research attention. PVA has become a popular material for disposable and biodegradable plastic alternatives because of these characteristics. For example, it could be used as a flocculant, metal-ion remover, and excipient in controlled release systems because of its water-soluble, biodegradable, and reactive properties [43].

3.3.2 Polyvinyl Acetate (PVAc)

A synthetic biodegradable polymer available in the form of a rubber-like compound is polyvinyl acetate (PVAc). It belongs to the family of polyvinyl esters and thermoplastic polymers in nature. For the synthesis of PVAc, vinyl acetate monomer undergoes free radical vinyl polymerization [44]. The chemical structure of polyvinyl acetate (PVAc) is shown in Fig. 11.

According to reports, PVAc biodegrades more slowly. In soil-burial testing, ethylene/vinyl acetate copolymers allow a relatively slow biodegradation process. Increasing the acetate content led to greater weight reduction for 120 days. There are several ways to manage PVAc's hydrolysis and subsequent oxidation, which should provide degradation materials with a wide variety of characteristics, degradability, and the ability to decompose in an environment-friendly manner [45].

It is widely used as glue-type material, generally referred to as carpenter's glue. In the paint industry, paper coating, and some other industrial coatings, and for nonwovens, sanitary napkins, and filter paper are used as a binder [46].

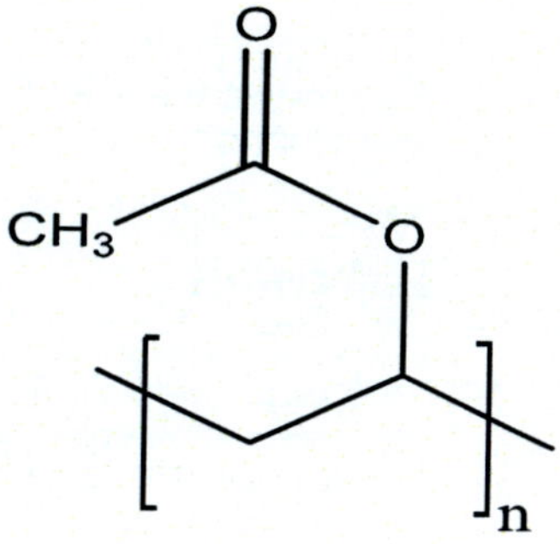

Fig. 11 Chemical structure of polyvinyl acetate (PVAc)

3.4 *Biodegradable Polyurethanes*

A diisocyanate, chain extender, and polyol combine to form polyurethanes. Hard and soft segments are present alternatively in a segmented polymer formed by the reaction of the two substances. Polyols like polyester polyols and polyether polyols are the building blocks for soft segments. With the help of the chain extender, a hard segment is synthesized from diisocyanate. Polyurethane biodegradation is influenced by the chemical composition of the polymer segments [47].

It is possible to control the degradation by selecting the right soft segment. Polyether-based polyurethanes are resistant to decomposition by microorganisms. Polyurethanes are quickly biodegradable if the polyol used is polyester. There are three types of biodegradable polyesters: PCL, PLA, and PGA, which are employed. When esters are placed in soft segments, it is considered that their breakdown rate is controlled. Hydrolysis is difficult for the urethane boundaries in the hard portion [48].

To satisfy the ever-changing needs of current technologies including coatings, adhesives, fiber, foams, and thermoplastic elastomers, the unique material polyurethane having a broad range of chemical and physical properties has been thoroughly adapted and tailored.

4 Conclusions

The ability to easily biodegrade towards the end of their service life is one of the most valuable advantages of biodegradable polymers over other types of polymers. But this property is limited and subjected to such conditions that facilitate their biological degradation. Biodegradable polymers have gained more attention during the last decades because of their prospective applications in the fields which are addressing environmental protection and sustainability. Currently, only a few synthetic-origin biopolymers have market acceptance because of a lack of price competitiveness with common synthetic polymers. To improve the different properties of biodegradable polymers, a lot of methods have been studied, developed, and applied accordingly. A promising future for the development of biopolymer materials is ahead.

References

1. Tiseo, I.: Production forecast of thermoplastics worldwide from 2025 to 2050 (2022)
2. Degnan, T.: Waste-plastic processing provides global challenges and opportunities. MRS Bull. **44**(6), 436–437 (2019). https://doi.org/10.1557/mrs.2019.133
3. Daria, M.: Landfill impacts on the environment—review. 1–16 (2019)
4. Verma, R., Vinoda, K.S., Papireddy, M., Gowda, A.N.S.: Toxic pollutants from plastic waste—a review. Procedia Environ. Sci. **35**, 701–708 (2016). https://doi.org/10.1016/j.proenv.2016.07.069
5. Thiounn, T., Smith, R.C.: Advances and approaches for chemical recycling of plastic waste. J. Polym. Sci. **58**(10), 1347–1364 (2020). https://doi.org/10.1002/pol.20190261
6. Cho, R.: The truth about bioplastics
7. Pillai, C.K.S.: Recent advances in biodegradable polymeric materials. Mater. Sci. Technol. (United Kingdom) **30**(5), 558–566 (2014). https://doi.org/10.1179/1743284713Y.0000000472
8. Leja, K., Lewandowicz, G.: Polymer biodegradation and biodegradable polymers—a review. Polish J. Environ. Stud. **19**(2), 255–266 (2010)
9. Arutchelvi, J., Sudhakar, M., Arkatkar, A., Doble, M., Bhaduri, S., Uppara, P.V.: Biodegradation of polyethylene and polypropylene. Indian J. Biotechnol. **7**(1), 9–22 (2008)
10. Lucas, N., Bienaime, C., Belloy, C., Queneudec, M., Silvestre, F., Nava-Saucedo, J.E.: Polymer biodegradation: mechanisms and estimation techniques—a review. Chemosphere **73**(4), 429–442 (2008). https://doi.org/10.1016/j.chemosphere.2008.06.064
11. Briassoulis, H.: The institutional complexity of environmental policy and planning problems: the example of Mediterranean desertification. J. Environ. Plan. Manag. **47**(1), 115–135 (2004). https://doi.org/10.1080/0964056042000189835
12. Jakubowicz, I.: Evaluation of degradability of biodegradable polyethylene (PE). Polym. Degrad. Stab. **80**(1), 39–43 (2003). https://doi.org/10.1016/S0141-3910(02)00380-4
13. Scott, G.: Photo-biodegradable plastics : their role in the protection of the environment. **29**, 135–154
14. Rabek, J.F.: Chapter 4. Oxidative degradation of polymers. Compr. Chem. Kinect. **14**(C), 425–538 (1975). https://doi.org/10.1016/S0069-8040(08)70336-4
15. Hofmann, D., Entrialgo-Castaño, M., Kratz, K., Lendlein, A.: Knowledge-based approach towards hydrolytic degradation of polymer-based biomaterials. Adv. Mater. **21**(32–33), 3237–3245 (2009). https://doi.org/10.1002/adma.200802213
16. Dreier, J., Brütting, C., Ruckdäschel, H., Altstädt, V., Bonten, C.: Investigation of the thermal and hydrolytic degradation of polylactide during autoclave foaming. Polymers (Basel). **13**(16) (2021). https://doi.org/10.3390/polym13162624
17. Vroman, I., Tighzert, L.: Biodegradable polymers. Materials (Basel) **2**(2), 307–344 (2009). https://doi.org/10.3390/ma2020307
18. Albertsson, A.C., Varma, I.K.: Aliphatic polyesters: synthesis, properties and applications. Adv. Polym. Sci. **157**, 1–40 (2002). https://doi.org/10.1007/3-540-45734-8_1
19. Douka, A., Vouyiouka, S., Papaspyridi, L.M., Papaspyrides, C.D.: A review on enzymatic polymerization to produce polycondensation polymers: the case of aliphatic polyesters, polyamides, and polyesteramides. Prog. Polym. Sci. **79**, 1–25 (2018). https://doi.org/10.1016/j.progpolymsci.2017.10.001
20. Avérous, L.: Polylactic acid: synthesis, properties and applications. Monomers Polym. Compos. Renew. Resour. 433–450 (2008). https://doi.org/10.1016/B978-0-08-045316-3.00021-1

21. Kaseem, M.: Properties and medical applications of polylactic acid related papers melt rheology of poly(Lactic Acid)/low density polyet ethylene polymer blends Mosab Kaseem preparation and characterization of binary and ternary blends with poly(Lactic Acid), pol [Online]. www.expresspolymlett.com
22. Siakeng, R., Jawaid, M., Ariffin, H., Sapuan, S.M., Asim, M., Saba, N.: Natural fiber reinforced polylactic acid composites: a review. Polym. Compos. **40**(2), 446–463 (2019). https://doi.org/10.1002/pc.24747
23. Nagarajan, V., Mohanty, A.K., Misra, M.: Perspective on polylactic acid (PLA) based sustainable materials for durable applications: focus on toughness and heat resistance. ACS Sustain. Chem. Eng. **4**(6), 2899–2916 (2016). https://doi.org/10.1021/acssuschemeng.6b00321
24. Lu, D.R., Xiao, C.M., Xu, S.J.: Starch-based completely biodegradable polymer materials. Express Polym. Lett. **3**(6), 366–375 (2009). https://doi.org/10.3144/expresspolymlett.2009.46
25. Samantaray, P.K., et al.: Poly(glycolic acid) (PGA): a versatile building block expanding high performance and sustainable bioplastic applications. Green Chem. **22**(13), 4055–4081 (2020). https://doi.org/10.1039/D0GC01394C
26. Li, Y., Liao, C., Tjong, S.C.: "Synthetic biodegradable aliphatic polyester nanocomposites reinforced with nanohydroxyapatite and/or graphene oxide for bone tissue engineering applications. Nanomaterials **9**(4) (2019). https://doi.org/10.3390/nano9040590
27. Gorth, D., Webster, T.J.: Matrices for tissue engineering and regenerative medicine. Biomater. Artif. Organs. 270–286. Elsevier (2011)
28. Labet, M., Thielemans, W.: Synthesis of polycaprolactone: a review. Chem. Soc. Rev. **38**(12), 3484–3504 (2009). https://doi.org/10.1039/b820162p
29. Guarino, V., Gentile, G., Sorrentino, L., Ambrosio, L.: Polycaprolactone: synthesis, properties, and applications (2017)
30. Christen, M.O., Vercesi, F.: Polycaprolactone: how a well-known and futuristic polymer has become an innovative collagen-stimulator in esthetics. Clin. Cosmet. Investig. Dermatol. **13**, 31–48 (2020). https://doi.org/10.2147/CCID.S229054
31. Niaounakis, M.: Definitions of terms and types of biopolymers (2015)
32. Aliotta, L., Seggiani, M., Lazzeri, A., Gigante, V., Cinelli, P.: A brief review of poly(Butylene Succinate) (PBS) and its main copolymers: synthesis, blends, composites, biodegradability, and applications. Polymers (Basel) **14**(4), 844 (2022). https://doi.org/10.3390/polym14040844
33. Rafiqah, S.A., et al.: A review on properties and application of bio-based poly(Butylene succinate). Polymers (Basel) **13**(9), 1–28 (2021). https://doi.org/10.3390/polym13091436
34. Stępień, K., et al.: Biocopolyesters of poly(butylene succinate) containing long-chain biobased glycol synthesized with heterogeneous titanium dioxide catalyst. ACS Sustain. Chem. Eng. **7**(12), 10623–10632 (2019). https://doi.org/10.1021/acssuschemeng.9b01191
35. Bondeson, D., Syre, P., Niska, K.O.: All cellulose nanocomposites produced by extrusion. J. Biobased Mater. Bioenergy **1**(3), 367–371 (2008). https://doi.org/10.1166/jbmb.2007.011
36. Ren, M., et al.: Crystallization kinetics and morphology of poly(butylene succinate-co-adipate). J. Polym. Sci. Part B Polym. Phys. **43**(22), 3231–3241 (2005). https://doi.org/10.1002/polb.20539
37. Nagata, M., Machida, T., Sakai, W., Tsutsumi, N.: Synthesis, characterization, and enzymatic degradation studies on novel network aliphatic polyesters. Macromolecules **31**(19), 6450–6454 (1998). https://doi.org/10.1021/ma980644f
38. Jian, J., Xiangbin, Z., Xianbo, H.: An overview on synthesis, properties, and applications of poly(butylene-adipate-co-terephthalate)–PBAT. Adv. Ind. Eng. Polym. Res. **3**(1), 19–26 (2020). https://doi.org/10.1016/j.aiepr.2020.01.001
39. Burford, T., Rieg, W., Madbouly, S.: Biodegradable poly(butylene adipate-co-terephthalate) (PBAT). Phys. Sci. Rev. (2021). https://doi.org/10.1515/-2020-0078
40. Monvisade, P., Loungvanidprapa, P.: Synthesis of poly(ethylene adipate) and poly(ethylene adipate-co-terephthalate) via ring-opening polymerization. Eur. Polym. J. **43**(8), 3408–3414 (2007). https://doi.org/10.1016/j.eurpolymj.2007.05.009
41. Havstad, M.R.: Biodegradable plastics. Plastic Waste and Recycling, pp. 97–129. Elsevier (2020)

42. Aslam, M., Kalyar, M.A., Raza, Z.A.: Polyvinyl alcohol: a review of research status and use of polyvinyl alcohol-based nanocomposites. Polym. Eng. Sci. **58**(12), 2119–2132 (2018). https://doi.org/10.1002/pen.24855
43. Gaaz, T., Sulong, A., Akhtar, M., Kadhum, A., Mohamad, A., Al-Amiery, A.: Properties and applications of polyvinyl alcohol, halloysite nanotubes and their nanocomposites. Molecules **20**(12), 22833–22847 (2015). https://doi.org/10.3390/molecules201219884
44. Emblem, A.: Plastics properties for packaging materials. Packaging Technology, pp. 287–309. Elsevier (2012)
45. Amann, M., Minge, O.: Biodegradability of poly(vinyl acetate) and related polymers, pp. 137–172
46. Kaboorani, A., Riedl, B.: Mechanical performance of polyvinyl acetate (PVA)-based biocomposites. Biocomposites, pp. 347–364. Elsevier (2015)
47. Kim, G.B., Guo, J., Hu, J., Shan, D., Yang, J.: Novel applications of urethane/urea chemistry in the field of biomaterials. Advances in Polyurethane Biomaterials, pp. 115–147. Elsevier (2016)
48. Tatai, L., et al.: Thermoplastic biodegradable polyurethanes: the effect of chain extender structure on properties and in-vitro degradation. Biomaterials **28**(36), 5407–5417 (2007). https://doi.org/10.1016/j.biomaterials.2007.08.035

Bioinspired Polymer Composite Implants

Ali Mehboob and Hassan Mehboob

Abstract Long bone fractures are treated with internal fixation prostheses such as screws, pins, intramedullary nails, and bone plates, depending on the type and nature of the fracture. Different prostheses exhibit dissimilar fixation constructs which are vital for callus generation and fracture bridging. Metallic implants have a significant mismatch with bone mechanical properties and create stress concentrations at the plate, which results in stress shielding. This phenomenon impedes load transmission at the fracture site, which can lead to non-union, bone mass loss, delayed healing, refracture, and construct failure. Flexible fiber-reinforced composite prostheses respond to biological friendly healing (secondary healing) and promote callus generation and soft tissue maturation and can provide solutions to problems. These polymer implants promote bioactivity around the implant. Further, the polymer composites biomechanical properties can be tuned easily by adding the functional powders into matrix and changing the type or direction of reinforcement fibers. The performance of these composites from the published work according to different materials is discussed. This chapter concludes that the bioinspired polymer composites have the potential to replace traditional metallic implants for orthopedic applications.

Keywords Orthopedic implants · Biomechanics · Bioactive polymer composites · Mechanical properties · Polymer matrix · Reinforcement fibers

A. Mehboob (✉)
Department of Textile Engineering, School of Engineering & Technology, National Textile University, Faisalabad, Pakistan
e-mail: shalimehboob@gmail.com; alimehboob@ntu.edu.pk

H. Mehboob
Department of Engineering Management, College of Engineering, Prince Sultan University, Riyadh, Kingdom of Saudi Arabia
e-mail: hmehboob@psu.edu.sa

K. Shaker and A. Hafeez (eds.), *Advanced Functional Polymers*, Engineering Materials,
https://doi.org/10.1007/978-981-99-0787-8_8

1 Introduction

Composites are defined as the materials that consist of two or more fundamentally different components to achieve the target properties superior to those of individual material used [1]. Usually there are two main parts of composites: matrix and reinforcements. Matrix is the base material, which acts as binder for the reinforcement fibers, transfers load, and defines the composite shape along with surface texture. While the reinforcements are used to enhance the mechanical properties of composites along the desired axis by changing the fibers' orientation [2]. Composites majorly can be distinguished into ceramic matric composites (CMCs), polymer matric composites (PMCs), and metal matric composites (MMCs) [3].

Bone fracture occurs when the strain limit of bone exceeds. Fractures mostly occur in the long bones and the causes of such fractures include injuries from vehicle mishaps, falls, sports, and sometimes high muscular forces [4]. Fractures disrupt the blood supply with damage to the external tissues. Depending on these unpredictable forces the nature of fractures varies which may result in the fractured bones pieces to be aligned correctly (stable fracture) or misaligned (displaced fracture). Thus, the fractures could be classified according to

- The position of the fracture (shaft portions: proximal, medial, and distal).
- Shapes of fractures (crack, diagonal, along the length or shaft fractures).
- Open fractures (muscles or skin torn) or closed fractures (no skin rupture).

A perfect implant must provide satisfactory initial healing and permit suitable load transfer to the fractured area. Implants such as bone plates, intramedullary nails, pins, and external fixators are used to fix the broken bones. Conventional implants are made up of metals displacing the poor biomechanical performance including bone resorption, fatigue degradation, implant failure, screw lag, X-ray artifacts, corrosive nature, and re-surgery [5]. Replacing materials should display the unique properties such as corrosion resistant, non-toxic, biocompatible, light weight, improved mechanical properties, and non-allergic as well. Advanced properties of polymer composites using reinforcement materials grabbed the researcher's attention. Recently, bio-composite implants are of great attention to replace these conventional metallic implants as the mechanical and biological properties can largely be altered [6]. Polymers are widely used as matrix and continuous synthetic or natural fibers (e.g., carbon fibers, glass fibers, bamboo fibers, etc.) as reinforcements for these biomedical implants [7]. Moreover, the surrounding tissue growth during their interaction with implants can also be enhanced due to antibacterial and bioactive nature of composite implants. That is why, these materials can be introduced as the replacement of broken bones or damaged parts and are named as biomaterials [8]. Such potential materials are used in dental and orthopedic implants, as their varying properties enable better biomechanical compatibility with bone tissues [9, 10].

Most challenging concerns to attain the required composite implants in medical are their fabrication methods. Various fabrication techniques are being followed by the researchers. The choice of proper fabrication methods to manufacture composites is one of the most challenging issues in medical science to achieve the desired implants. Traditional fabrication techniques including hand layup, gas foaming, phase separation, compression molding, hot extrusion, solvent casting, melt mixing, additive manufacturing, and much more [11]. These techniques have their own advantages and disadvantages but still in the developing phase by the researchers to maximize the performance with reduced drawbacks. This chapter covers the biomechanical aspects of composite implants, their manufacturing techniques, performance, and future aspects.

2 Nature of Bone Healing

Once the fracture occurs, all the emergency repairing cells in blood report the injury site and start the repairing procedure. Depending on the fracture's nature, location, and biomechanical environment, the repairing procedure follows the specific healing mechanism [12] as follows:

1. Primary healing (intramembranous ossification).
2. Secondary healing (endochondral ossification).

In primary healing, direct bone formation occurs where the mesenchymal cells directly differentiate into osteoblast cells and blood vessels growth without the callus formation at fracture site. Secondary healing is the common form of bone fracture healing with the process of callus formation as shown in Fig. 1. This type of healing follows the four main stages:

i. Hematoma formation:

   This is an inflammatory action where the blood clots at the fracture site as a result of the local cells' death due to traumatic condition of blood vessels' rupture. This process continues for first week which is the initial stage of calluses formation.

ii. Fibro-cartilaginous callus formation:

   This period continues for second and third week after fracture occurrence. In this preliminary stage, the soft calluses define their shapes wrapping the fracture site to stabilize the broken bones according to the available biomechanical environment. The mesenchymal cells are able to differentiate into fibroblast, chondrocytes, and osteoblasts which are responsible for the production of soft tissues, cartilaginous tissue, and bony tissues, correspondingly. The fibrous and cartilaginous tissues are more prominent with some portions of woven bones in calluses.

iii. Bony callus formation:

   In this step, the woven bone is developed into mature lamellar bone. This process continues from first till fourth month in which almost whole fracture

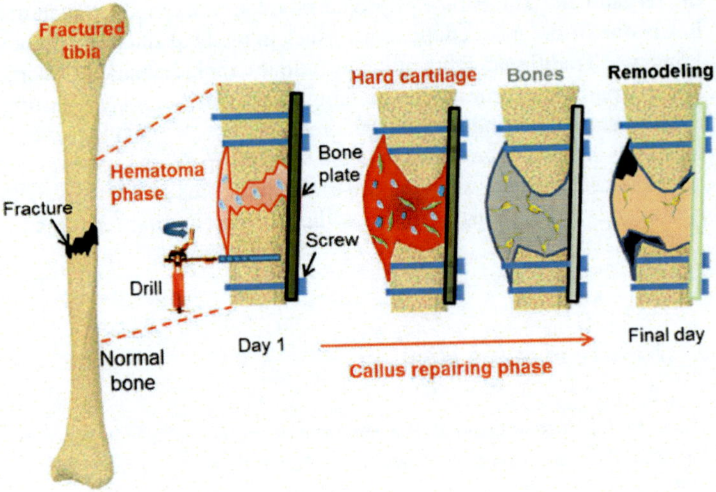

Fig. 1 Callus formation during fracture healing

stabilizes the fractured bones and starts bearing the body weight. The stresses are reduced from the implant assembly and exchanged with the calluses.

iv. Resorption phase:

In this process, the developed bone starts to dissolve itself in the physiological environment and restores its original shape. This process continues for months and years [13].

Depending on these healing mechanisms and fracture's location, the fractures are majorly divided into two types:

(a) Articular fractures:

The fractures occur at the proximal and distal end of the long bones or the bones where trabecular bone is greater than the cortical bone. These fractures are generally minor or hair line fractures. Rigid fixation techniques are used to fix these fractures disgracing any movement. Direct bone formation takes place following the primary healing.

(b) Diaphyseal fractures:

Diaphysis (cylindrical shaft) of the long bones is the main candidate which faces such fractures where the cortical bone is the main runner. For the effecting healing of these fractures, the implant should provide sufficient stability to broken bones while allowing the micro-movement to promote the improved stimulus in calluses. This will allow the gradual sharing of body weight with stress relief at the implant-bones assembly.

3 Why Composites?

Metallic implants highly mismatch the material properties as that of bones resulting in the aforementioned complications, especially erosion during long-term remains [14]. An ideal implant should provide reasonable initial micro-movement at fracture site to stimulate healing and permit suitable load transfer to the fractured area. The metallurgical composition of metallic implants was enhanced with improved corrosion resistance and edges design were presented in 1909 and afterward in 1912; these bone plates because of poor fatigue properties were abandoned. The following improvement was made in 1948, when a plate with sliding vertical pockets was created for the screws assembly to prevent stress concentration [15]. Due to insufficient structural stability, improvement in this plate design was also required. In 1965, a new plate design was presented with a fixing tensioner that permitted the interfragmentary development of broken bones. This development process continued along with the introduction of locking compression plate and dynamic compression plate but still there are many controversial theories about the satisfactory performance of metallic implants. Composite materials grabbed the attention of research due to their enhanced features as shown in Fig. 2. The preferred mechanical properties for special fracture cases can be achieved by altering the stacking sequence of the fibrous composites, while keeping the basic structure the same [16]. Elastic composite implants provide proper stability along with the favorable biomechanical environment to promote the secondary healing process and reduce the stress shielding phenomenon. This process increases the callus volume which envelops the fractured area causing the bony bridging to facilitate bone-implant stability and decrease the risk of implant failure [17]. In the last decade, one of the extensively studied areas is bio-composite materials for medical applications. Therefore, biodegradable, bioactive, antibacterial, and nanomaterial FRP composites are strong candidates to meet the biomechanical requirements in the biomedical surgery as orthopedic implants (for knee, jaw, elbow, hip, leg, ribs, and dental) to collaborate with the biomechanics of bones [18, 19].

4 Materials for Composite Implants

Large range of materials are used these days to achieve the target properties of implants for desired applications. The implant materials should be biocompatible means the material which acts as an appropriate host with minimum disturbance in the body function. These materials include metals, ceramics, powders, polymers, fibers, and their composites [20, 21] while their existence period and decomposition behavior define if they are non-degradable, partially degradable, fully degradable, biocompatible, or bioactive. These properties tend to differ in mechanical properties of these materials which are listed in Table. 1. The industry has grown over 150

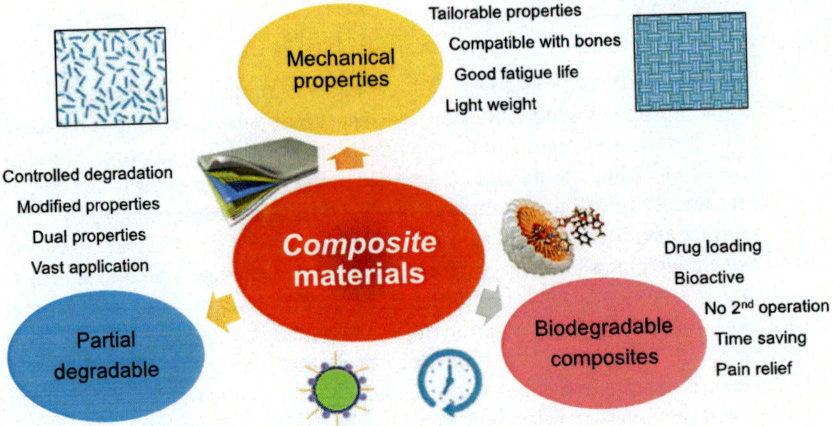

Fig. 2. Composites for implants application

billion US dollars [22], serving the quality life to millions of patients by treating their fractured bones.

4.1 Non-degradable Materials

Non-degradable materials are those which maintain their integrity both mechanically and physically. Traditionally non-degradable metallic materials were massively used for the implants manufacturing. Advancement in artificial polymers technologies in recent years resulted in various medical devices and were effectively used for the treatment of patients around the world. These

materials include stainless steel (SS), titanium (Ti), cobalt (Co), chromium (Cr), silver (Ag), and platinum (Pt) [23] while synthetic polymers include poly(ethylene), poly(propylene), poly(methyl methacrylate), poly(dimethyl siloxane), poly(ether ketone), and polyurethane [24].

Manufacturing of implants began with pure metals which displayed reduced strength and high corrosion. Improved mechanical- and corrosion-resistant properties of stainless steel brought an innovation in market as they are stiff, ductile, and dense. But release of metal ions and other chemical compounds caused material deterioration with pitting and stress corrosion resulting in irritation and allergic reaction in the body [25, 26]. Titanium is light weight, higher torque, and fatigue resistant as compared to stainless steel [27]. Moreover, titanium implants are better corrosion resistant as oxide layer regenerates quickly. The high metal mechanical properties cause stress shielding along the bone resulting in bone resorption and mass loss. Various methods were introduced to reduce the structural stiffness of implant like in bone plate the higher density was maintained around the fracture and reduced

Table 1 Material properties of biomaterial implants

	Material	Material properties	Degradation properties	Medical applications
		Young's modulus (GPa)	Time (months)	
Lower limb long bones	Cortical bone (longitudinal dir.)	15–20	–	–
	Cortical bone (transverse dir.)	8.5	–	–
	Cortical bone (hoop dir.)	6.9	–	–
	Trabecular bone	1.1	–	–
Tissue phenotype in bone healing process	Granulation tissue	0.00002	–	–
	Fibrous tissue	0.001	–	–
	Cartilage	0.005–0.5	–	–
	Immature bone	0.5–1.0	–	–
	Intermediate bone	1.0–2.0	–	–
	Mature bone	2.0–6.0	–	–
Metals	Stainless steel	190	Years	Bone plates, IM rods, screws, joint replacements, tissue engineering, dental implants, heart valves, stent spinal implants
	Co-Cr alloys	210–235	Years	
	Titanium	110	Years	
	Titanium alloy (Ti-6Al-4V)	116	Years	
	Magnesium alloys (AZ91, AZ31, WE43, Mg-Ca, LAE442, Mg-Mn-Zn)	45	Months ~ Years	
Ceramics	Alumina	380	–	Orthopedic prosthesis, bone fillers, dental implants, jawbone reconstructions, facial surgery, ear, nose and throat repair
	Zirconia	150–220	–	
	Calcium phosphate (HA, TCP)	35–110	–	
Bioglass	45S5	35	0.6–1	–
	13–93	60	0.6–1	–
Polymers	UHMWPE (ultra-high molecular weight polyethylene)	12~4	–	Bone plates, IM rods, screws, disks, dental implants, prosthesis, spine cage, suture anchors, meniscus repair
	PEEK (Polyether ether ketone)	3.6	–	
	Polysulfones (PS)	2.6	–	
	Polyethylene terephthalate (PET)	2.8	–	
	Polylactic acid (PLA)	0.3–2	18~12	

(continued)

Table 1 (continued)

	Material	Material properties	Degradation properties	Medical applications
	Polyglycolic acid (PGA)	0.7–5	0.4–3	
	Poly(lactic-co-glycolic acid) (PLGA)	1.4–2.8	0.6–2	
	Polycaprolactone (PCL)	0.4–0.6	>24	
Polymer composites	CF/PEEK, GF/PEEK, CF/epoxy	10–70	Months ~ Years	–
	BG/polylactide, polyglycolide and its copolymers			–
	CF/PS, BG/PU, BG/PS, CF/UHMWPE			–

moving toward the edges [28]. Attention was drawn for the further improvement using metallic alloys like Co-Cr (cobalt, chromium, and traces of molybdenum) and Ti-Co. Cr promotes oxide formation on the surface which decreases corrosion while molybdenum discourages bulk corrosion [29]. In artificial hip implants, Co-based alloys are used due to better ductility with improved wearing resistance during physiological loading conditions where high-strength applications are required [30]. Titanium was then introduced as implant material with biocompatible, light weight, and mechanical properties closer to bone making it better option over others. It forms titanium oxide as corrosion-resistant layer on the surface.

Polymer-based composites have more fatigue resistance compared to other materials with high strength and low stiffness. Carbon fiber-reinforced polymer (CFRP) composites are potential candidates for orthopedic applications [31]. Monitoring the implant designing factors in vivo and in vitro is a time-taking process with involvement of various ethical issues. Thus, introducing the computer models to check the effect of implants' performance is made much easier to tune the various designing parameters and their effect on fracture healing. It was concluded that these CFRP composites can effectively reduce the stress concentration in the bone [32, 33, 16, 34]. Improvement in structure was done by sandwiching the flax/epoxy between glass/epoxy composite. This hybrid composite provided flexible inner surrounded with stiffer shell providing more ideal conditions for fixing bone fractures [35]. Only the mechanical performance of the implants was investigated while the biological response was not considered. In vivo tests need to be done for their biomechanical performance validation as a safe implant.

Moving toward the synthetic polymers, chemical linkages made them resistant to degradation in the human body and were assessed broadly for medical applications but few displayed satisfactory results as implants. Major interest of using polymers was its altering mechanical properties by changing the reinforcement orientation. Ultra-high molecular weight poly(ethylene) (UHMWPE) being a tough material, resistant to wear, chemically stable, less friction, and reduced moisture permeability made them to dominate over four decades starting from 1962 as joint replacement material [36]. Large number of 1.4 million patients were treated during this period. However, long-term biological exposure reduced their mechanical performance which led to re-surgery. Investigation of this failure exposed the requirement of altering the fabrication process like mixing with additional materials, changing cross-linkage which improved the performance including surface properties [37]. Polypropylene (PP) was used as woven mesh to reinforce the implants but showed inflammatory response leading to pain [38]. Introduction of foreign materials during surgery is critical and the complete healing process can take 12 weeks. Thus, various materials were used as coatings or blends to improve the biomechanical response of the implants facilitating the bone repair.

4.2 Partially Degradable Materials

Partially degradable materials are those which display two behaviors at the same time, one is non-degradable part and the other is degradable. Non-degradable part maintains its physical, mechanical, and chemical integrity throughout its stay while degradable portion initially helps in the improvement of mechanical properties and then starts to degrade as time progress. The target of the implant is to provide the initial stability, but their mechanical properties should cooperate with bones. As the material starts to degrade, the bioactivity starts which enables the cell adhesion, stimulus to embryonic tissues, and antibacterial properties as well. These materials could be applied as coatings on the implant surfaces, blend during fabrication process, or fillers in the porous structures. Materials for surface coatings of implants include hydroxyapatite, calcium phosphate, polylactic acid (PLA), polylactic-co-glycolide acid (PLGA), silver, MXene, magnesium, zinc oxide nanoparticles, and iodine [35, 39–41]. These coatings start to dissolve as they encounter the body fluid and perform their duties by releasing ions and nutrients. Further, to decrease the structural modulus of the solid implants for target application, various techniques were introduced like 3D printing (additive manufacturing) of porous structure [42–44]. These structures can be tuned by changing their cellular structure (cubic, honeycomb, diamond, circular, and body-centered cubic) along with the pore size, shape, and strut thickness simply by modeling software [45]. These porous structures manufactured using metallic biomaterials can be filled using degradable polymers which are safe and broadly used in medical industry [46]. Recently, titanium bone plates with varying porosities of cubic cellular structures were prepared and the cavities were filled with polyglycolic acid (PGA). Titanium is responsible of the elastic–plastic properties as a major load carrier while PGA degraded till seventh week losing its maximum properties [47]. Thus, the partially degradable materials are multifunctional with better performance as of non-degradable materials.

4.3 Fully Degradable Materials

These are the materials that are completely resorbable with the passage of time into the body. The concept of these implants came due to the various complications with the materials discussed previously. Once the calluses heal properly, we go for the second surgery to remove the conventional metallic implants which will cause pain, disruption of soft tissues, blood loss, and sometimes infection. As the fracture heals, the calluses start sharing the load with the implant used to fix the broken bones. With the passage of time, the calluses are mature enough to withstand the body weight. Here is the point when implant should start losing its strength during the healing process with the load-sharing mechanism. This was the motivation of developing the biodegradable composites which will dissolve in the physiological environment once the fracture heals. It will diminish all conventional problems of metallic implants. Majorly, under clinical applications, the biodegradable materials are categorized into three groups, (a) metals, (b) bioceramics, and (c) polymers, and combinations of these materials as composites.

(a) Recently, unidirectional magnesium alloy wires have been used to reinforce PLA for in vitro tests [48]. Magnesium alloys [49] were not invented for medical use but their biocompatible attitude and excellent mechanical properties made them a suitable candidate to be used as fixation devices which are comparable with those of human cortical bone. Thus, various alloys were developed for these applications such as Mg-Ca, Mg-Zn, and Mg-Al. Magnesium devices were introduced as bars and rods [50] at 5–20% volume content, and wires in uni-directional or braided form at 6–10% fraction at different angles.

(b) Ceramics are well known because of their biocompatibility, bioactivity, and corrosion resistance behavior. But the drawbacks are their brittleness, complex shapes, high modulus, and poor fatigue. Thus, the ceramics are used as reinforcement materials in polymers as they alone are not feasible for medical devices. Calcium phosphate, tri-calcium phosphate, and hydroxyapatite are well known bioceramics.

(c) The origin of the natural polymers is starch, collagen, silk, alginate, and chitosan and synthetic polymers are polylactic acid, polyglycolic acid, poly-l-lactic acid (PLLA), polycaprolactone (PCL), polydioxanone (PDS), and polylactic co-glycolic acid (PLGA). The use of these polymers alone cannot be used because of their low strength.

Thus, polylactic acid (PLA) has been studied mostly because of its excellent biocompatible and biodegradable properties. To make the PLA compatible with the requirement of load-bearing bone fractures, various reinforcements are required. Its modulus and tensile strength have been described to be approximately 3.5 GPa and 59 MPa, correspondingly, which are much lower to bear the body weight [51]. As the reinforcement materials, Mehboob et al. [52] used unidirectional plasma-treated bioactive glass (BG) fibers as they offer excellent biomechanical properties. 30-second plasma treated displayed the elastic modulus of around 27 GPa with 118

MPa strength. But the BGF/PLA showed fast degradation when exposed to body fluid and thus magnesium (AZ31) alloy wires were used for reinforcement in PLA. The mechanical tests were carried out with varying volume content of 20–50%. These wires showed better sustainability in the human body [6]. Composite with 50 % volume content of Mg wires was selected for surface treatment of Mg wires for improved sustainability and stifled mass loss, stable pH, and encouraged the deposition of Ca and P compounds with better mechanical properties [6]. Further, research is going on to improve the biomechanical properties of these fully biodegradable composite implants to make them bioactive and antibacterial.

5 Implant Types and Their Applications

Treatment of broken bones is concerned not only with the repair of the bone anatomy but to consider the complete bone union with full recovery and increased callus volume. Callus is the soft tissue that envelops the fracture during early stabilization and converts into bone as time progresses. Enhanced callus volume ensures the better quality of healing. Various implants are used during operation to stabilize with minimum additional reduced injury to bone and surrounding soft tissues. These implants are majorly categorized into external (traction and pins) and internal devices (pins, screws, intramedullary nails, rods, and bone plates) for the treatment of orthopedic and dentistry fractures (see Fig. 3). Further the implants could be static or dynamic. Static implants are stiff and tightly screwed with bone without any relative motion while the dynamic implants are flexible around the assembly or with less stiff material providing the micro-movement at the fracture site. These devices are used according to the fracture location and its nature which decides the quality of bone union. Internal fixation devices are of main concern which are discussed in detail.

5.1 Screws

Orthopedic screws (OS) are usually used as fixation devices to fix the bone plates, intramedullary nails, and treat the minor bone fractures independently. The screws geometry has four parts: head, thread, shaft, and tip. For tightening the screw, screwdriver is attached to the head. This head could be hexagonal, circular, slotted, and straight. Head produces counterforce when compression is generated during fastening any implant to the bone. The shaft is the smooth surface between the threaded area and the head. The threads consist of the core diameter, thread diameter, pitch, and the lead with which screw advances during single complete turn into the bone. Thread core determines the contact area of the bone and screw resistance during pullout force. The screw tip is either fluted or round depending upon the way of insertion during tightening. Types of screws vary according to the length, its thread diameter, the distance between threads, and the design of their head and tip.

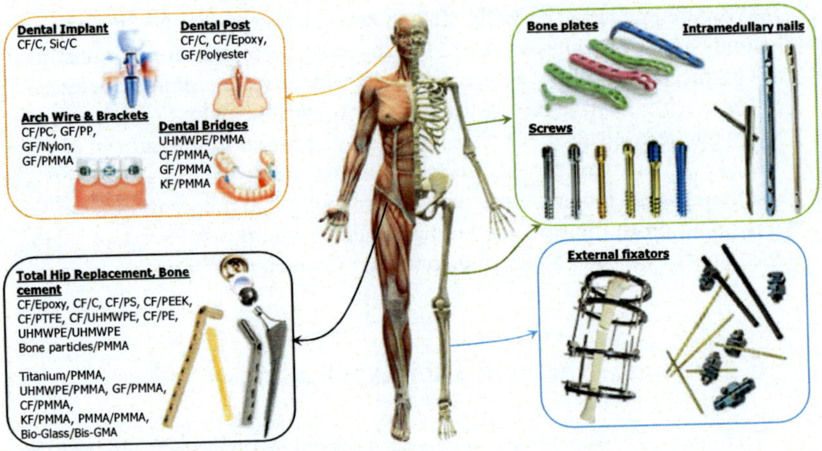

Fig. 3 Implants for fracture treatment

Based on these parameters and application, they are divided for the use of cortical and trabecular bone. For cortical screws, if the length of screw is very small and able to be tightened only in single cortical wall called as uni-cortical while the screws which hold both cortical walls when tightened are bi-cortical screws. These screws have dull ends with close spaced shallow threads overall the length [53, 54]. Mostly the holes are drilled before the insertion of these screws. The packing like the washer is placed if there is no bone plate between the screw head and the cortical bone to prevent any bone damage due to stress concentration. For trabecular bone, the screw threads are wide with large area to improve screw grip into the bone. Cannulated screws are hollow fixed using guide wire for better alignment to hold the multiple fracture segments instead of drilling. These are the basic types of screws used to repair small fractures.

5.2 *Bone Plates*

Bone plates are used to stabilize the broken bones by fixing them on the outer surface of the bone. These plates are available in different sizes, length, width, and shapes depending on the area of application. When fracture occurs, two different forces are produced on both sides of fractured bone: tensile and compressive forces. Pauwels defined the way of fracture fixation using tension band procedure for first time. During this principle, the bone plate is applied on the tension side of the fractured bone which neutralizes these tensile forces and translate them into compression forces. If the plate is directly applied on the compression side, it will bend or break due to fatigue. Plates help the primary function of muscles and joints by fixing the broken bones but early weight bearing is not recommended. Plates promote the risk of refracture after removal due to irritation, mass loss, and osteoporosis below the plate [55]. A rigid compression plate (CP) was first used in 1970 to fix the fractures and bone union was observed without callus formation promoting primary healing [56]. This fixation requires the plate compression against the bone relying on the tightening torque by the screw heads. CPs demonstrate disadvantages including: bending the plate before insertion to sit on the bone surface, damage of the one-third of vascular tissues, increased bone porosity around assembly, and mainly the stress shielding effect between the bone–bone plate [57]. To reduce the soft tissue injury, researcher tried to improve the plating technique by limiting the contact area of the bone plate–bone. Locking plate provides stability with locked screw heads into the plate maintaining the distance between bone plate–bone instead of compression with bone. To enhance the performance, flexible plates were prepared using vitallium alloys [58]. Still the controversies exist in using the completely rigid or flexible plates. It is believed that the micro-movement at the fracture site encourages fast fracture union due to the enlarged callus wrapping with combined primary and secondary healing process [59]. Bone plate rigidity depends on the material properties and cross-sectional area of the bone plate. That is why biomaterials with reduced stiffness and improved biomechanical performance implants are ruling the orthopedic industry by replacing the metallic implants [60]. Depending upon the fixation method and fracture's nature, plates are broadly grouped into four types: compression plates, buttress plates, bridge plate, and neutralization plates.

Compression plates as explained above are designed to fix fractures using compression. These compression plates are commonly referred to as the dynamic compression plates (DCPs). The holes in DCPs have tapered edges and the screw tights the plate against the bone without leaving any play.

Buttress plates are used to hold the fractures in place rigidly mostly at the proximal and distal ends of the long bones like at the ankle or knee where the fracture faces large forces. These plates are mostly wide to increase the strength. As these plates are used around complex surfaces, thus designed into L-shaped, T-shaped, or rounded ends.

Bridge plates as clear by the name are used to stabilize the comminuted fractures or large bone defects. Alone the function of this plate is very hard to stabilize the fracture and therefore scaffold-type artificial structure or bone grafts need to be placed between the fragments to relief the bone plate.

Neutralization plates are general plates where neutralization indicates that how the plate works during fracture fixation. This plate reduces the loading phenomenon on the fracture site by caring the loads and transferring them across the plate instead of the fracture site. Neutralization plates are fixed with screws to neutralize the torsional force, bending force, and shear forces. Wedge- or butterfly-type fractures are commonly fixed using these plates.

5.3 *Intramedullary Nails*

Intramedullary (IM) nails have gained attention and accepted universally over more than past 50 years. The nails are inserted into the medullary canals of the long bones following the hammering (HP) or reaming procedures (RP) assembled with screws at the proximal and distal ends of bone. During hammering, the nail press fits with the bone into the canal by the frictional forces and secures the movement of separated bone fragments. This fixation method is preferred for the mid-shaft diaphyseal long bone fractures especially in lower limb. The advantage is that the insertion method of IM nails requires small incision without extensive surgical procedure. They are in contact at multiple points of longitudinal bone interface while during reaming, a reamer is used to drill a coaxial hole along the bone length into the medullary canal. Gap is created around 1mm between the IM nail and bone interface promoting the flexible fixation. HP is more stable during axial and rotational distortion of fracture while more damage is done to the bone marrow and trabecular bone with RP. An ideal intramedullary nail is still under development with improvement in design and material although the varying bone anatomy makes it impossible. Better implant should have the following conditions:

It should provide enough stability by maintaining the alignment and position with reduced rotational movement as the axial compression is recommended.
It should provide the needed biomechanical stimulus for bone union.
It should be positioned in a way that its removal is facilitated after bone union.

Fracture stability with torsional and bending stiffness in nailing mainly depends on the material properties, diameter of the medullary canal, number and types of screws, and the diameter of the nail.

6 Fabrication Techniques

Numerous fabrication methods are used to prepare bioactive composites for fracture fixations. Commonly used fabrication processes include hot pressing, additive manufacturing, melt extrusion, injection molding, and compression molding. Various forms of implants with desired shapes can be fabricated including scaffolds, films, disks, screws, intramedullary nails, and bone plates.

6.1 *Melt Extrusion*

Melt extrusion in early 1930s was introduced in manufacturing industries. This technique was then used for the fabrication of orthopedic composite implants. During melt extrusion, raw material is melted and forced into a die of the desired shape.

Various steps include: (1) heating the material until it melts, (2) mixing of desired materials to obtain the homogeneous composite blend, and (3) applying pressure using piston or screw to the prepared material into the die for desired composite shape. These advantages include the following:

Cost-effective procedure.
Various material combination.
Solvent absence reduced toxicity.
Elimination of time taking drying process.
Complex shapes achievable.

While melt extrusion also reveal some disadvantages. High energy systems are required for heating chambers, push drives and only 2D shapes can be achieved. Degradable materials including ceramics, metals, polymers, and glasses in powder form can be easily processed with polymer pellets or fibers to enhance the biomechanical properties of composite implants.

6.2 *Hot Compression Molding*

In this technique, various composite materials including several combinations of the long and short fibers, aligned and random-oriented fibers, powders, and ceramics can be fabricated. Films of the pure polymers or in combination with powders are prepared using hot compression into the mold. Then the reinforcements prepared in the form of mats, weaved or random-oriented fibers, are placed in between the polymer films during staking sequence. Once the staked layers are placed in the mold, the mold is preheated up to the glass-transition temperature and the pressure is applied to the upper moveable surface of the mold to compress the stacked layer for even distribution of matrix into the reinforcement materials removing all possible trapped air. Then the mold is left at room temperature to cool down the fabricated composite. Different shapes with varying volume fraction and fibers orientation can be achieved using this fabrication technique.

6.3 *Injection Molding*

This process is same as that of melt extrusion process. The main difference is the addition of mold in which the melted material is injected to the cavity of the enclosed mold with desired shape. The material is then allowed to cool by rapid heat removal for composite solidification. This process is automatic and the even distribution of matric cannot be controlled.

6.4 Additive Manufacturing

Additive manufacturing (AM) is a technique that is grabbing much attention in recent years for three-dimensional (3D) printing of material. This technique was brought to the medical industry for biomaterial printing. This technique includes various methods like the fused deposition modeling (FDM), selective laser sintering (SLS), bioprinting, direct metal laser sintering (DMLS), and stereolithography (SL).

Fused deposition modeling is widely used for 3D printing mainly for the polymer materials along with their composites reinforced with chopped and continuous fiber. The materials used for FDM include thermoplastics including PLA or PLLA along with their blends such as carbon fibers, Kevlar, glass fibers, and onyx. The part is prepared using melted filament of thermoplastic extruded from the printer heated nozzle in semi-solid form to lay up the layer-by-layer structure. This technique aims high resolution with brilliant accuracy compared to other AM techniques.

Selective laser sintering is another 3D printing technology which was first introduced in 1980s. The high performance of SLS includes superior accuracy, resolution with smooth surface finish compared to other 3D plastic printing technologies. Any kind of shape can be produced using high power laser beam by fusing the small powder or polymer particles. The remaining unfused powder supports the printed structure without any additional structure attached. These printed parts have good mechanical properties compared to other 3D printing technologies along with low production cost with fast manufacturing making it trendy in AM.

7 Future Directions; Challenges and Opportunities

Degradable implants should meet two biomechanical criteria: one is initial integrity with degraded mechanical compatibility during healing period and second is the biocompatibility and bioactivity between the implant and the host bone fracture. Therefore, these two criteria should be considered while designing a biodegradable implant. The first objective is to provide the adequate initial mechanical support to the fractured bones followed by optimal gradual degradation. During this period of bone healing, the biodegradable implant provides temporary structural support for tissue differentiation process and completely degrades in a certain time frame, which depend on the severity of fracture, fracture site, fracture geometry, age, and weight. Generally, an orthopedic implant should provide the mechanical stability from 1 to 6 months in lower limb long bone fractures. However, it remains a challenge to set the balance between implant mechanical properties and healing rate of fractured bones. Therefore, careful considerations should be given to the material of implant and fracture-related parameters influence the healing of bone fractures. Moreover, a reliable in vivo or in vitro degradation study should be conducted according to the FDA recommendations following ISO-10993 standard if biodegradable implant is designed. Therefore, a biodegradable implant and the clinical implications such as healing of bone fractures should be critically investigated. Additional functions such as bioactive materials, drug delivery mechanism, and bone healing monitoring via smart biosensors should be integrated into the biodegradable implants. Appropriate bioactivity, optimal amount of drug release, and the biosensors for real-time monitoring of healing status need more advancements and extensive investigations are required to design the next-generation biodegradable orthopedic implants.

References

1. Evans, S.L., Gregson, P.J.: Composite technology in load-bearing orthopaedic implants. Biomaterials **19**, 1329–1342 (1998). https://doi.org/10.1016/S0142-9612(97)00217-2
2. Mehboob, H., Chang, S.H.: Application of composites to orthopedic prostheses for effective bone healing: a review. Compos. Struct. **118**, 328–341 (2014). https://doi.org/10.1016/J.COMPSTRUCT.2014.07.052
3. Shue, B., Moreira, A., Flowers, G.: Review of recent developments in composite material for aerospace applications. Proc. ASME Des. Eng. Tech. Conf. **1**, 811–819 (2010). https://doi.org/10.1115/DETC2009-87847
4. Court-Brown, C.M., McBirnie, J.: The epidemiology of tibial fractures. 77, 417–421 (1995). https://doi.org/10.1302/0301-620X.77B3.7744927
5. Dickson, K., Katzman, S., Delgado, E., Contreras, D.: Delayed unions and nonunions of open tibial fractures. Clin. Orthop. Relat. Res. 302:189–193 (1994). https://doi.org/10.1097/00003086-199405000-00029
6. Ali, W., Mehboob, A., Han, M.G., Chang, S.H.: Experimental study on degradation of mechanical properties of biodegradable magnesium alloy (AZ31) wires/poly(lactic acid) composite for bone fracture healing applications. Compos. Struct. **210**, 914–921 (2019). https://doi.org/10.1016/J.COMPSTRUCT.2018.12.011
7. Zhao, R., Su, H., Chen, X., Yu, Y.: Commercially available materials selection in sustainable design: an integrated multi-attribute decision making approach. Sustain 8, 79 8:79 (2016). https://doi.org/10.3390/SU8010079
8. Shackelford, J.F.: Bioceramics—current status and future trends. Mater. Sci. Forum 293, 99 (1999). https://doi.org/10.4028/WWW.SCIENTIFIC.NET/MSF.293.99
9. Dental implant complications: etiology, prevention, and treatment—google books. https://books.google.com.pk/books?id=QecbCQAAQBAJ&pg=PA231&lpg=PA231&dq=Klein,+M.O.;+Schiegnitz,+E.;+Al-Nawas,+B.+Systematic+review+on+success+of+narrow-diameter+dental+implants.Int.+J.+Oral+Maxillofac.+Implants2014,29,+43-5&source=bl&ots=uW9Ac0L34P&sig=ACfU3U0ZiXh-B2WkLgFVe-73I55pE6tvyg&hl=en&sa=X&ved=2ahUKEwiZvYXxqvz2AhVCXBoKHcPtBgcQ6AF6BAglEAM#v=onepage&q=Klein%2C. Klein, M.O., Schiegnitz, E., Al-Nawas, B.: Systematic review on success of narrow-diameter dental implants. Int. J. Oral Maxillofac. Implants. Accessed 5 Apr 2022
10. Silva, V.V., Lameiras, F.S., Domingues, R.Z.: Microstructural and mechanical study of zirconia-hydroxyapatite (ZH) composite ceramics for biomedical applications. Compos. Sci. Technol. **61**, 301–310 (2001). https://doi.org/10.1016/S0266-3538(00)00222-0
11. Reddy Nagavally, R.: Composite materials-history, types, fabrication techniques, advantages, and applications (2016)
12. Claes, L., Augat, P., Suger, G., Wilke, H.J.: Influence of size and stability of the osteotomy gap on the success of fracture healing. J. Orthop. Res. **15**, 577–584 (1997). https://doi.org/10.1002/JOR.1100150414
13. Mehboob, H., Kim, J., Mehboob, A., Chang, S.H.: How post-operative rehabilitation exercises influence the healing process of radial bone shaft fractures fixed by a composite bone plate. Compos. Struct. **159**, 307–315 (2017). https://doi.org/10.1016/J.COMPSTRUCT.2016.09.081

14. Fouad, H.: Effects of the bone-plate material and the presence of a gap between the fractured bone and plate on the predicted stresses at the fractured bone. Med. Eng. Phys. **32**, 783–789 (2010). https://doi.org/10.1016/J.MEDENGPHY.2010.05.003
15. Uhthoff, H.K., Poitras, P., Backman, D.S.: Internal plate fixation of fractures: short history and recent developments. J. Orthop. Sci. **11**, 118–126 (2006). https://doi.org/10.1007/S00776-005-0984-7
16. Mehboob, A., Mehboob, H., Chang, S.H.: Evaluation of unidirectional BGF/PLA and Mg/PLA biodegradable composites bone plates-scaffolds assembly for critical segmental fractures healing. Compos. Part A Appl. Sci. Manuf. **135**, 105929 (2020). https://doi.org/10.1016/J.COMPOSITESA.2020.105929
17. Mehboob, A., Chang, S.H.: Effect of composite bone plates on callus generation and healing of fractured tibia with different screw configurations. Compos. Sci. Technol. **167**, 96–105 (2018). https://doi.org/10.1016/J.COMPSCITECH.2018.07.039
18. Auclair-Daigle, C., Bureau, M.N., Legoux, J.G., Yahia, L.: Bioactive hydroxyapatite coatings on polymer composites for orthopedic implants. J. Biomed. Mater. Res. Part A **73A**, 398–408 (2005). https://doi.org/10.1002/JBM.A.30284
19. Hussain, M., Khan, S.M., Al-Khaled, K., et al.: Performance analysis of biodegradable materials for orthopedic applications. Mater. Today Commun. **31**, 103167 (2022). https://doi.org/10.1016/J.MTCOMM.2022.103167
20. Krishnakumar, S., Senthilvelan, T.: Polymer composites in dentistry and orthopedic applications-a review. Mater. Today Proc. **46**, 9707–9713 (2021). https://doi.org/10.1016/J.MATPR.2020.08.463
21. Prakasam, M., Locs, J., Salma-Ancane, K., et al.: Functional biomaterials review biodegradable materials and metallic implants-a review (2017).https://doi.org/10.3390/jfb8040044
22. Liu, Q., Jiang, L., Shi, R., Zhang, L.: Synthesis, preparation, in vitro degradation, and application of novel degradable bioelastomers—a review. Prog. Polym. Sci. **37**, 715–765 (2012). https://doi.org/10.1016/J.PROGPOLYMSCI.2011.11.001
23. Davis, R., Singh, A., Jackson, M.J., et al.: A comprehensive review on metallic implant biomaterials and their subtractive manufacturing. Int. J. Adv. Manuf. Technol. **2022**, 1–58 (2022). https://doi.org/10.1007/S00170-022-08770-8
24. Gunatillake, P.A., Adhikari, R.: Nondegradable synthetic polymers for medical devices and implants. Biosynthetic Polym. Med. Appl. 33–62. https://doi.org/10.1016/B978-1-78242-105-4.00002-X
25. Chen, Q., Thouas, G.A.: Metallic implant biomaterials. Mater Sci. Eng. R Rep. **87**, 1–57 (2015). https://doi.org/10.1016/J.MSER.2014.10.001
26. Eliaz, N.: Corrosion of metallic biomaterials: a review. Materials (Basel) 12 (2019). https://doi.org/10.3390/MA12030407
27. Hayes, J.S., Richards, R.G.: The use of titanium and stainless steel in fracture fixation. Expert Rev. Med. Devices **7**, 843–853 (2010). https://doi.org/10.1586/ERD.10.53
28. Ramakrishna, K., Sridhar, I., Sivashanker, S., et al.: Design of fracture fixation plate for necessary and sufficient bone stress shielding. JSME Int. J., Ser. C Mech. Syst. Mach. Elem. Manuf. **47**, 1086–1094 (2004). https://doi.org/10.1299/JSMEC.47.1086
29. Khan, W., Muntimadugu, E., Jaffe, M., Domb, A.J.: Implantable medical devices, pp. 33–59 (2014). https://doi.org/10.1007/978-1-4614-9434-8_2
30. Pandey, A., Awasthi, A., Saxena, K.K.: Advances in materials and processing technologies. ISSN: (Print) (Online) Journal homepage: https://www.tandfonline.com/loi/tmpt20. Metallic implants with properties and latest production techniques: a review Metallic implants with properties and latest production techniques: a review (2020). https://doi.org/10.1080/2374068X.2020.1731236
31. Tayton, K., Johnson-Nurse Brian Mckibbin, C., Bradley, J., Hastings, G.: The use of semi-rigid carbon-fibre-reinforced plastic plates for fixation of human fractures results of preliminary trials. From the department of traumatic and orthopaedic surgery, Cardiff Royal infirmary, and North Staffordshire polytechnic, Stoke-on-Trent (1982)

32. Mehboob, A., Mehboob, H., Chang, S.H., Tarlochan, F.: Effect of composite intramedullary nails (IM) on healing of long bone fractures by means of reamed and unreamed methods. Compos. Struct. **167**, 76–87 (2017). https://doi.org/10.1016/j.compstruct.2017.01.076
33. Mehboob, A., Mehboob, H., Kim, J., et al.: Influence of initial biomechanical environment provided by fibrous composite intramedullary nails on bone fracture healing. Compos. Struct. **175**, 123–134 (2017). https://doi.org/10.1016/j.compstruct.2017.05.013
34. Mehboob, H., Chang, S.H.: Evaluation of healing performance of biodegradable composite bone plates for a simulated fractured tibia model by finite element analysis. Compos. Struct. **111**, 193–204 (2014). https://doi.org/10.1016/J.COMPSTRUCT.2013.12.013
35. Manteghi, S., Mahboob, Z., Fawaz, Z., Bougherara, H.: Investigation of the mechanical properties and failure modes of hybrid natural fiber composites for potential bone fracture fixation plates. J. Mech. Behav. Biomed. Mater. **65**, 306–316 (2017). https://doi.org/10.1016/J.JMBBM.2016.08.035
36. Costa, L., Luda, M.P., Trossarelli, L.: Ultra-high molecular weight polyethylene: I. Mechano-oxidative degradation. Polym. Degrad. Stab. **55**, 329–338 (1997). https://doi.org/10.1016/S0141-3910(96)00170-X
37. Sobieraj, M.C., Rimnac, C.M.: Ultra high molecular weight polyethylene: mechanics, morphology, and clinical behavior. J. Mech. Behav. Biomed. Mater. **2**, 433–443 (2008). https://doi.org/10.1016/j.jmbbm.2008.12.006
38. Meintjes, J., Yan, S., Zhou, L., et al.: Synthetic, biological and composite scaffolds for abdominal wall reconstruction. Expert Rev. Med. Devices **8**, 275–288 (2011). https://doi.org/10.1586/ERD.10.64
39. Shaker, K., Ashraf, M., Jabbar, M., et al.: Bioactive woven flax-based composites: development and characterisation. 46, 549–561 (2016).https://doi.org/10.1177/1528083715591579
40. Alt, V.: Antimicrobial coated implants in trauma and orthopaedics–a clinical review and risk-benefit analysis. Injury **48**, 599–607 (2017). https://doi.org/10.1016/J.INJURY.2016.12.011
41. Rasool, K., Helal, M., Ali, A., et al.: Antibacterial Activity of Ti3C2Tx MXene. ACS Nano 10 (2016). https://doi.org/10.1021/acsnano.6b00181
42. Amin Yavari, S., Ahmadi, S.M., Wauthle, R., et al.: Relationship between unit cell type and porosity and the fatigue behavior of selective laser melted meta-biomaterials. J. Mech. Behav. Biomed. Mater. **43**, 91–100 (2015). https://doi.org/10.1016/J.JMBBM.2014.12.015
43. Chakraborty, A., Datta, P., Majumder, S., et al.: Finite element and experimental analysis to select patient's bone condition specific porous dental implant, fabricated using additive manufacturing. Comput. Biol. Med. **124**, 103839 (2020). https://doi.org/10.1016/J.COMPBIOMED.2020.103839
44. Mehboob, H., Tarlochan, F., Mehboob, A., et al.: A novel design, analysis and 3D printing of Ti-6Al-4V alloy bio-inspired porous femoral stem. J. Mater. Sci. Mater. Med. **31**, 1–14 (2020). https://doi.org/10.1007/S10856-020-06420-7/TABLES/3
45. Mehboob, H., Tarlochan, F., Mehboob, A., Chang, S.H.: Finite element modelling and characterization of 3D cellular microstructures for the design of a cementless biomimetic porous hip stem. Mater. Des. **149**, 101–112 (2018). https://doi.org/10.1016/J.MATDES.2018.04.002
46. Gohil, S.V., Suhail, S., Rose, J., et al.: Polymers and composites for orthopedic applications. Mater. Devices Bone Disord. 349–403 (2017). https://doi.org/10.1016/B978-0-12-802792-9.00008-2
47. Mehboob, H., Mehboob, A., Abbassi, F., et al.: Finite element analysis of biodegradable Ti/polyglycolic acid composite bone plates based on 3D printing concept. Compos. Struct. **289**, 115521 (2022). https://doi.org/10.1016/J.COMPSTRUCT.2022.115521
48. Li, X., Chu, C.L., Liu, L., et al.: Biodegradable poly-lactic acid based-composite reinforced unidirectionally with high-strength magnesium alloy wires. Biomaterials **49**, 135–144 (2015). https://doi.org/10.1016/J.BIOMATERIALS.2015.01.060
49. Sheikh, Z., Najeeb, S., Khurshid, Z., et al.: Biodegradable materials for bone repair and tissue engineering applications. Materials (Basel) **8**, 5744–5794 (2015). https://doi.org/10.3390/ma8095273

50. Staiger, M.P., Pietak, A.M., Huadmai, J., Dias, G.: Magnesium and its alloys as orthopedic biomaterials: a review. Biomaterials **27**, 1728–1734 (2006). https://doi.org/10.1016/j.biomaterials.2005.10.003
51. Jung, K.-C., Han, M.-G., Mehboob, A., Chang, S.-H.: Performance evaluation of bone plates consisted of BGF/PLA composite material according to body fluid exposure conditions. Compos. Res. **30**, 21–25 (2017). https://doi.org/10.7234/COMPOSRES.2017.30.1.021
52. Mehboob, H., Bae, J.H., Han, M.G., Chang, S.H.: Effect of air plasma treatment on mechanical properties of bioactive composites for medical application: composite preparation and characterization. Compos. Struct. **143**, 23–32 (2016). https://doi.org/10.1016/J.COMPSTRUCT.2016.02.012
53. Diagnostic imaging evaluation of the postoperative patient following musculoskeletal trauma. Semantic Scholar. https://www.semanticscholar.org/paper/Diagnostic-imaging-evaluation-of-the-postoperative-Weissman-Reilly/bc0ab6ddcb84ed8dabb3a71dc771a6b00e957249. Accessed 10 April 2022
54. Richardson, M.L., Kilcoyne, R.F., Mayo, K.A., et al.: Radiographic evaluation of modern orthopedic fixation devices. 7:685–701 (1987). https://doi.org/10.1148/RADIOGRAPHICS.7.4.3329363
55. Daugherty, K., Burns, B.: Campbell's operative orthopaedics, 11th edn. Mosby/Elsevier, Philadelphia PA (2008)
56. Allgöwer, M., Perren, S., Matter, P.: A new plate for internal fixation—the dynamic compression plate (DCP). Injury **2**, 40–47 (1970). https://doi.org/10.1016/S0020-1383(70)80111-5
57. Wagner, M.: General principles for the clinical use of the LCP. Injury **34**, 31–42 (2003). https://doi.org/10.1016/J.INJURY.2003.09.023
58. Claes, L.: The mechanical and morphological properties of bone beneath internal fixation plates of differing rigidity. J. Orthop. Res. **7**, 170–177 (1989). https://doi.org/10.1002/JOR.1100070203
59. Perren, S.M.: Evolution of the internal fixation of long bone fractures. The scientific basis of biological internal fixation: choosing a new balance between stability and biology. J. Bone Joint Surg. Br. **84**, 1093–1110 (2002). https://doi.org/10.1302/0301-620X.84B8.13752
60. Mccartney, W., Mac Donald, B.J., Hashmi, M.S.J.: Comparative performance of a flexible fixation implant to a rigid implant in static and repetitive incremental loading. J. Mater. Process. Technol. **169**, 476–484 (2005). https://doi.org/10.1016/J.JMATPROTEC.2005.04.104

Functional Polymers and Composites for Zero Gravity

Khubab Shaker, Yasir Nawab, and Madeha Jabbar

Abstract The current chapter focuses on the functional polymer and composite materials used for zero gravity applications. It discusses the various structures that are used in zero gravity (zero-G) and the material requirements for these structures. A structure in zero-G experiences harsh environmental conditions including the damaging effects of high vacuum, atomic oxygen (ATOX), radiations (ultraviolet and ionizing radiation), micrometeorites (i.e., space debris), and thermal cycling. The commonly used materials for zero gravity include polymers for thermal blankets or electronic components, adhesives, polymer aerogels, shape memory polymers, fiber reinforced composite materials, fiber metal laminates, protective coatings against atomic oxygen exposure and lunar dust adhesion, etc. All these materials have been detailed in this chapter and concluded with future trends in the domain.

Keywords Zero-G · Shape memory polymers · Fiber reinforced composites · Aerogels · Testing · Surface engineering

1 What Is Zero Gravity?

The term zero gravity is used for the state or condition when no apparent force of gravity is acting on a body and is abbreviated as zero-G. As the distance of any object from the center of earth increases, the gravitational force acting on it decreases. Newtonian mechanics defines the weight of a body as the force exerted on it due to gravitational pull. Therefore, the situation in which no gravitational forces are acting on a body is also called a state of weightlessness. Bodies undergoing free fall and

K. Shaker (✉)
Department of Materials, School of Engineering and Technology, National Textile University, Faisalabad, Pakistan
e-mail: khubab@ntu.edu.pk

Y. Nawab · M. Jabbar
Department of Textile Engineering, School of Engineering and Technology, National Textile University, Faisalabad, Pakistan

K. Shaker and A. Hafeez (eds.), *Advanced Functional Polymers*, Engineering Materials,
https://doi.org/10.1007/978-981-99-0787-8_9

those in orbit experience this situation. For example, manmade satellites in orbit around the earth experience zero gravity when they are on board.

1.1 Systems Used in Zero Gravity

The three main components of every space program are the satellite, the launch vehicle, and the space center. A satellite is the most common manmade system used in zero gravity, orbiting around Earth or any other body in space [1]. The very first satellite (Sputnik) was launched into space by the Soviet Union in 1957. Since then, almost 9000 satellites have been launched by more than 40 countries. The satellites provide a birds-eye view of the earth, allowing us to see large areas and collect more data at one time, as compared to terrestrial instruments. Therefore, satellites were initially used for military and defense purposes. However, they are being used widely for communication (television, telephone, etc.), navigation, weather and climate monitoring, space science, etc. [2].

The increased dependency on satellites has led to the exponential growth of this industry. The trend has moved toward smaller satellites, owing to their compact size, reduced cost, and development time. An additional benefit is the reduced launch costs and elimination of expensive propulsion systems. These small satellites are usually in Low Earth Orbit (LEO), with an altitude range from 250 to 1000 miles (not more than about one-third of the radius of Earth). Satellites in LEO can also be effectively used for satellite communications. Global communications have become more robust with LEO as signal time delays are of the order of only 5–10 ms, and it takes 90 min to orbit [3].

There are 2,666 active satellites in orbit, 1,918 of which are in low Earth orbit (LEO), as of April 2020 statistics on active satellites in space. The global small satellite market worth was 3.22 billion USD in 2020 and is projected to grow at a cumulative annual growth rate (CAGR) of 16.4% to reach 13.71 billion USD by 2030 [4]. Other systems used in zero-G include spacecraft destined to explore the Solar System, e.g., those used for lunar probes, planetary probes (Mars, Mercury, Jupiter, etc.), solar observation missions, space probes, etc. However, the focus of this chapter is on the zero-G systems in LEO, i.e., satellites.

1.2 Environment in LEO

A small satellite in LEO (zero-G) undergoes harsh environmental conditions. These include the damaging effects of high vacuum, ATOX (atomic oxygen), radiations (ultraviolet and ionizing radiations), space debris (termed micrometeorites), and thermal cycling [5]. The effect of these parameters (summarized in Table 1) plays a critical part in determining the reliability, function, and life of the system. The atmosphere of Earth is thick, consisting of gases mainly nitrogen (78%) and oxygen

Table 1 Effect of environmental conditions on zero-G systems [6]

Environment	Effect
High vacuum	Volatilization (for materials having low vapor pressure), vacuum welding, diffusion
Atomic oxygen (ATOX)	Oxidation, surface erosion, crazing/cracking of materials
Radiations (ultraviolet and ionizing radiation)	Develops lattice defects, polymer chain scission, cross-linking (for certain organic materials)
Micrometeorites (space debris)	Material fracture, mechanical failure
Thermal cycling	Embrittlement, chemical, and mechanical degradation

(21%), and eliminates much of the space radiation. However, the satellite in LEO also experiences atmospheric drag, causing it to move to lower orbits and eventually fall to the earth.

In extraterrestrial environments (Moon, Mars, etc.), dust is also a critical obstacle for space missions. The dust particulates (diameter $\leq$ 60 μm) are abrasive, electrostatically charged, chemically reactive, and have a tendency to adhere to exposed surfaces. About 50% of these dust particles come from the lunar regolith. During the Apollo missions, lunar dust posed unexpected difficulties such as visor, glove, and boot abrasion, deterioration of seals and thermal radiator function, and respiratory irritation [7].

1.3 Material Requirements for Satellites

Keeping in view the environment in LEO, the construction material used for the satellite should meet the following requirements [8].

a. Ability to work in a hard vacuum.
b. Erosion resistance against atomic oxygen.
c. Endurance over temperature extremes (−150 to 60 °C).
d. Prevent surrounding contamination (very low outgassing).
e. Resistance to harsh ultraviolet and ionizing radiations.

A material may or may not need to satisfy all these requirements simultaneously, depending on its application area. For example, a material directly exposed to the surface of spacecraft should satisfy all these, but those used on the inner side may have some exceptions.

2 Materials Used for Zero-G Applications

The various materials used include metals and their alloys, polymers, and composite materials. The metallic materials commonly used include aluminum, magnesium, iron, and titanium, and their properties are compared in Table 2.

However, the chapter mainly focuses on polymers and composite materials used for zero-G applications, and the detail of metallic materials is beyond the scope of this chapter.

2.1 *Polymers in Space*

Polymeric materials are used for numerous applications in a space system including adhesives, thermal coatings, paints, thermal insulations, seals, tapes, toughening/damping materials, thin film substrates, potting compounds, etc. The polymers for these applications are designed to last for 15–20 years, depending on the system life. Consequently, it is necessary to comprehend how space conditions affect the polymeric materials now in use [6]. Additionally, the polymeric material is qualified for space application only if it meets certain requirements, defined as per end use. Materials that might corrode the metal surface, cause an electrical short circuit, contaminate optics, or adversely affect other components are not used for these applications [8].

Polymers are either organic or inorganic in nature. The inorganic materials cannot experience loads, while organic materials have poor resistance to heat. The optimal structure must be made of a straight-arranged polymer with nonionic substituent clusters to enable adaptability in order to produce a material between these constraints. The success of inorganic polymers (commonly silicone polymers) is based on this fundamental notion [10]. Some typical applications of polymeric materials for space structures include their use in thermal blankets, heat control paints, and adhesives.

Table 2 Comparison of properties of Al, Mg, Fe, and Ti [9]

Property	Fe	Mg	Al	Ti
Yield strength (MPa)	280	90–195	~170	434
Density (kg/m^3)	7900	1700	2700	4600
Strength-to-weight ratio	35	114.7	62.9	94.3
Elongation (%)	12–45	14–45	5–25	18–30
Modulus of elasticity (GPa)	196–207	45	70	107–119
Melting point (°C)	1538	650	660	1668
Thermal expansion coefficient ($\times 10^{-5}$)	1.18	2.48	2.31	0.86
Thermal conductivity (W/m K)	79	160	235	22

2.1.1 Thermal Blankets

Thermal blankets are used in space structures to regulate their temperature, offering a stead range of operating temperature. It consists of a polymeric film, either filled with carbon black to absorb the sunlight or coated with a vapor deposited aluminum layer to reflect the sunlight. Multiple layers of these polymeric films (separated by fine nylon cloth) are used to constitute a multi-layer insulation termed as thermal blanket. The most commonly used polymers for films include polyethylene terephthalate) and polyimide (commercially available from DuPont). Additionally, a thin layer (thickness 50 nm) of indium tin oxide is deposited on polymeric films to provide a path for static electricity dissipation.

2.1.2 Paints

The paints are a key component of the structure to regulate the temperature of spacecraft. It consists of a pigment dispersed in some organic or inorganic binder. The paints are either black or white in color depending on the material used. The black paints have tendency to absorb the sunlight and keep the spacecraft at warmer temperatures, while white paints reject excess heat back into space, and have high emissivity. The polyurethanes are used for black paints, while silicones are used for white paints.

2.1.3 Adhesives

Adhesives are preferred in space structures for durable bonding of dissimilar materials, allowing significant mass reduction as compared to conventional mechanical fixation compared to traditional screws or rivets. Typical applications of adhesive include structural bonding, thread-locking compounds (to avoid loosening of fasteners under high vibration), lamination of optical elements, and cable stacking. But a key requirement is that adhesive shall withstand vibrational loads due to the spacecraft launch and must be able to withstand the impact of space environment [11].

The adhesive families commonly preferred for spaces programs include epoxies (for better mechanical performance), silicones (for better performance in thermal environment), acrylics (used as tapes), and polyurethanes. The two-part epoxy is the most widely used adhesive, in which two liquid components are mixed, and allowed to cure into a tough, and adherent solid. The properties of epoxy are modified by incorporating fillers, to get adhesives with low thermal expansion coefficient, exceptional adhesion, and high temperature stability (to 300 °C).

Other adhesives include the ones used for specialty applications, i.e., electrical conductivity, thermal conductivity, etc. The specialty grade adhesives are filled with silver or carbon powder to impart electrical conductivity, or boron nitride, alumina, and diamond dust to provide thermal coupling. Good adhesion and intimate contact are the only ways to achieve electrical or thermal continuity between uneven surfaces.

Electronic component spot bonding can also be done with conductive adhesives. The UV curing transparent adhesives constitute another class of specialty adhesives. Such adhesives are preferred for bonding together the optical elements, such as lenses and fiber optic devices.

2.1.4 Polymers for Electronic Components

Polymers are widely used in electronic components and for electronic packagings. Some of the applications include conductive adhesives, circuit boards, conformal coatings, wire insulation, etc. The circuit boards are based on polymer impregnated glass cloth. Some typical examples include an epoxy impregnated glass cloth and poly(cyanate) resin impregnated quartz cloth. For wire insulation, Teflon is preferred due to its chemical inertness; however, it has poor radiation stability. A recent development is the use of poly vinylidene fluoride (PVDF) polymer for cable insulations, used in high radiation environment. Circuit boards are coated with conformal coatings to prevent chemical deterioration and add an insulating layer to the surface. Mostly these coatings are a two-part urethane that is blended, dipped, or sprayed onto circuit boards.

2.2 *Shape Memory Polymers*

Shape memory polymers (SMP) are known as the stimuli-responsive polymeric materials. External stimuli, like light, temperature, electric current, etc. have the ability to cause SMP to return to their original structure. These polymers comprise a network of molecular switching segments, i.e., the hard segments and soft segments. The hard segments define the permanent shape of SMP, while soft segments become less rigid in response to a specific stimulus. This phenomenon allows to program the polymer into a temporary form. When exposed to a certain stimulus, the molecular switches are activated, releasing the stored strain energy that causes the temporary form to take on its original permanent shape. The temperature is a common external stimulus since most polymers have glass transition temperature (T_g) or melting temperature (T_m) [12]. The SMP when heated over this threshold temperature return to its permanent shape. This critical temperature is often the glass transition temperature of SMP.

Shape memory polymers offer advantages including low cost, lightweight, high storage energy ability, and recovery in response to a specific stimulus. Moreover, SMP have high strength-to-weight ratio as compared to metallic materials, as well as flexibility of design. Therefore, SMP have the potential to replace metallic components in spacecraft. Epoxy-resin is one of the prominent members of SMP family. The shape memory effect and superior performance properties (high modulus, heat resistance, creep resistance, better adhesion, electrical resistance, etc.) of epoxy-resin promote its use for deployable space applications [13].

The shape memory polymer composites (SMPC) reinforced with fibrous reinforcement show significant improvement in the mechanical properties and durability as compared to SMP. The shape memory effect of SMPC does not degrade considerably due to UV and γ radiation. A shelter or habitat is utilized to shield space travelers from the harsh conditions of space when they spend longer periods of time in outer space. Such habitats are constructed on earth and transported to space. Deployable structures are developed in order to maximize deployed volume in space while minimizing occupied space in spacecraft. One of the lightweight materials that can fulfill the need is SMPC since it can be released in space using a specific stimulus while being packaged and stowed in a lower volume [14].

2.3 *Polymer Aerogels*

Aerogels are ultralight polymeric materials in which the liquid component of a gel is replaced with a gas. These polymer aerogels can be used as solvent absorbents, super-insulators, or catalyst supports. Aerogels made from hydrophobic, semicrystalline polymers, have high stiffness combined with excellent flexibility [15]. The properties of aerogel depend on its chemical composition, density, and synthesis route followed. Aerogels may be created in a variety of forms, including monoliths, grains, powders, and films. The monolithic silica aerogels were used by NASA for space applications as thermal insulation material.

The structure of an aerogel is made up of small amount of silica particles tortuously locked in a 3D network with numerous dead-ends that impede its thermal transport. The lattice structure, density, and interconnectivity of particles affect how much heat is transferred through the solid structure of aerogel.

Because of the interparticle connections, silica aerogels are often weak and brittle, rendering them unsuitable for load-bearing applications. Three-point bending, uniaxial compression, and ultrasonic techniques are common ways to characterize the mechanical performance of silica aerogel. Researchers are focusing on the improvement of the mechanical strength by the incorporation of a second phase (i.e., fibers) in these aerogels. It has been reported that elastic strength and modulus are increased by 26 and 85%, respectively, by addition of 10% of fibers. Incorporation of 5% carbon nanofibers into lattice structure improved the tensile strength and compressive modulus of aerogels by 5 and 3 times, respectively.

2.4 *Fiber Reinforced Polymer Composites*

The fiber reinforced polymer composites are widely used for aerospace applications. The commonly preferred composites for space applications include carbon fiber reinforced polymer (CFRP) laminates, while some other types are also used [16].

Table 3 Reinforcing fibers commonly used for aerospace structures [17]

Reinforcement	Strength (MPa)	Modulus (GPa)	Application area
Glass			
– E-glass	2200–2600	65–75	Radome, rocket motor casing, secondary parts
– S-Glass	4400–4800	85–95	High load bearing parts in small passenger aircrafts
Aramid			
– Low modulus	2700–2800	80–85	Fairings, non-load bearing parts
– High modulus	2300–2400	160–170	High load bearing parts
Carbon			
– High strength	3000–3500	220–240	Satellites, dish antenna, missiles, etc.
– High modulus	2800–3000	390–450	Space structures, control surfaces

The reinforcement materials preferred for space applications are summarized in Table 3.

FRPC used for aerospace are generally fabricated from prepreg as raw materials using an autoclave fabrication technique [18]. Sometimes, the filament winding approach is also prevalent for tubular/shell-like components, i.e., rocket motor casings used for launch vehicles and missiles. Generally, such parts are post-cured at room temperature, or an oven is used for this purpose. Resin transfer molding technique is also used to fabricate special components like radomes. The examples of some popular matrix systems are summarized in Table 4, along with their attributes.

Composite materials have long been employed in space applications, and their popularity is growing. Some of the typical applications include space flight vehicles, satellites, and launch vehicles (pressure vessels for fuel, rocket motors, etc.). The low weight and environmental stability of composite materials makes them an idea material for these applications. Composite materials are also preferred for ablative and high temperature components in reentry heat shields and motors [19].

The carbon fiber reinforced polymer (CFRP) laminates are one of the most abundantly used composite materials for aerospace applications, due to the high modulus of carbon fiber [20]. The ability to bear high temperatures and low thermal expansion are the additional benefits of using carbon fiber. The CFRP is widely used on satellites and payload support structures. A typical bus structure is built from aluminum honeycomb, sandwiched between CRFP face sheets. The CFRP is preferred for structures demanding dimensional stability over extreme temperatures and is produced with low moisture absorption resins. Common CRFP applications include RF reflectors, solar array substrates, struts and booms for deployable instruments, deployable payload fairings of launch vehicles, etc. Silica or carbon fiber reinforced phenolic high temperature resistant ablative composites are typically used, and they absorb heat via changing state [21]. These composites are utilized in rocket nozzles (throats

Table 4 Matrix materials commonly used for space structures [17]

Thermosets		Thermoplastics
Epoxy	Phenol	PPEK, PPS
Most widely used (80% of total)	Cheap, low viscosity, and easy to use	High mechanical strength and good damage tolerance
Density: 1.1–1.4 g/cc	Density: 1.2–1.4 g/cc	Density: 1.3–1.4 g/cc
Strength: 40–85 MPa	Strength: 35–60 MPa	Strength: 100 MPa
Modulus: 2.7–5.5 GPa	Modulus: 2.7–4.1 GPa	Modulus: 3.5–4.4 GPa
Moisture absorption: 5–6%, causes degradation of high-temperature properties	Absorbs moisture but does not affect performance at high temperature	No moisture absorption
Good storage to make prepreg	Difficult to prepreg and store	Difficult to process, 300–400 °C temperature is required
Low shrinkage: 2–3%	High cure shrinkage	
No volatiles during process	Volatiles release during process	Very high chemical resistance
Moderately high-temperature resistance	Preferred for high-temperature applications	Creep resistance
Degradation due to UV exposure		

and exit cones), reentry vehicle heat shields, etc., and are categorized into ablatives and ceramics.

Lopez et al. [22] prepared ultra-thin CFRP using POSS nanofiller as reinforcement in cycloaliphatic epoxy matrix. These composite materials were developed for use in Low Earth Orbit (LEO), offering enhanced protection against exposure to atomic oxygen. The addition of 20% POSS demonstrated the lowest erosion yield, which amounted to a time period of 12 months in a simulated low-earth orbit environment. Unidirectional CFRP is used in housing, boom, strap elements, and solar panels of LEO satellite, owing to their anisotropic tailorable properties.

The polybenzoxazine matrix has high concentration of hydrogen and is responsible for shielding against galactic cosmic rays. Additionally, because of its low polymerization temperature, the reinforcing fibers are not harmed during the fabrication of composite material. The low viscosity of this resin allows the use of sophisticated fabrication techniques like resin infusion for composite fabrication. In terms of combined structural and radiation shielding properties, the UHMWPE reinforced composites outperform the benchmark materials [23]. According to another research, glass and Kevlar fiber composites are used in the fabrication of satellite antennas because of their high electrical conductivity and minimal transmission loss.

Natural fiber reinforced composites (NFRC) also find application in aerospace structures [24]. The flax/phenol composites were the first to be used in main spar of the Bristol Breinheim bomber. Currently natural fiber sandwich panels (flax/phenol as skin and Nomex honeycomb as core) are used in aircraft interior as ceiling, flooring, walls, lavatories, liners, etc. Compression molding is the most preferable approach for the fabrication of these sandwich panels, using prepregs, at suitable curing cycle [25].

Shape memory polymer composites (SMPC), also called as elastic memory composites (EMC), are widely used for deployable space structures. These materials are similar to the conventional FRPC, except that the matrix material used is a thermosetting shape memory polymer [26]. As discussed in earlier section, SMP exhibit the shape memory effect as a result of some stimuli. Some examples of SMPC-based structural components in aerospace include deployable panels, solar arrays, morphing structures, reflector antennas, expandable lunar habitat, mandrels, etc. [27]. Solar arrays are the energy generation systems in a space structure and are commonly packed in the vehicle before launch. The solar array is deployed to gather energy once in space. The antenna is a crucial component in satellite-to-Earth communication because it may provide essential data about space-related issues. New forms of morphing wing skin that may alter the shape in response to temperature or other environmental stimuli are also being developed by researchers.

A unique habitat is required for prolonged space adventures in order to support basic life and protect against UV radiation harm. A deployable space habitat increases the work volume in space while minimizing the volume occupied in the spacecraft. It can be launched utilizing certain stimulus in space and bundled and stored in a smaller container when not in use. Water-soluble mandrels and multi-piece mandrels are used while fabricating complex structures. However, SMPC mandrels have the benefits of temperature-dependent stiffness variation, simple demolding, reusability, and lower product cost. However, SMPC mandrels offer the advantages of variable stiffness based on external temperature, easy demolding, reusable, and a lower product cost.

2.5 *Fiber Metal Laminates*

The fiber metal laminates (FMLs) are defined as the panels produced by sandwiching reinforcement in between metal sheets. This combination gives a synergistic effect, incubating properties of metal and reinforcement such as damage tolerance, corrosion resistance, fatigue endurance, thermal insulation, weight reduction, specific strength, and cost-effectiveness [28]. Owing to these advantages, FMLs have proven effective in Satellites, Launch Vehicles, and Space Centers application. The most common fibrous reinforcements used for FMLs include carbon, glass, and aramid fibers. These are abbreviated as CARAL (carbon-reinforced aluminum laminate), ARAL (aramid-reinforced aluminum laminate), and GLARE (glass-reinforced aluminum laminate) [29, 30]. Hybrid reinforcements containing a combination of glass, carbon, aramid, or other fibers have also been explored by researchers [31].

Patil et al. [32] reviewed the characterization of glass/epoxy reinforced aluminum laminates. They reported that although the carbon/epoxy reinforced aluminum laminates exhibit superior mechanical performance as compared to their glass fiber counterparts, they have an issue of galvanic corrosion and unstable interface that can be addressed by treatment and processing of aluminum. Bienias et al. [33] discussed the impact resistance and damage growth in CARAL. They attributed the excellent interfacial strength and damage tolerance to the synergistic effect of stiffness and good mechanical properties of Carbon-epoxy reinforcement along with the ductile behavior of aluminum. The best ply sequence reported by the authors was 0°/90° and ±45° orientation.

One of the most crucial design considerations for space applications is to lighten a structure without sacrificing its rigidity and strength. Due to their low weight, high specific bending stiffness, strength under dispersed loads, and strong energy-absorbing capabilities, honeycomb sandwich structures have been extensively employed in the production of aircraft structures. The honeycombs produced by laying up CFRP on aluminum sheets help to reduce weight by up to 33% as compared to their metallic counterparts. Boudjemai et al. [34] analyzed the hexagonal honeycomb panels used for satellite structures, both experimentally and numerically. They concluded that the geometry and material have a direct effect on the performance of honeycomb structure.

Dinca et al. [35] reported that in comparison to GLARE, CARAL has superior damage tolerance, fatigue, fracture resistance, tensile and bending strength. These laminates are an appealing choice for satellite structure applications due to their tailorable qualities via ply orientation, good damage tolerance, fatigue endurance, corrosion resistance, lightweight, and high specific strength. However, the practical application of these materials needs excessive research in their thermal and vibrational properties for qualification of the material system.

2.6 Lunar Dust Adhesion Mitigation Materials

Lunar dust is the fine proportion of regolith, present on the Moon's surface. The lunar dust might have a negative impact on human outpost technology and crew personnel. For example, the surfaces might be damaged by the abrasive nature of dust, and the friction may have an adverse effect on coatings as well as wiring. The dust can potentially damage astronaut lungs, and cardiovascular systems, and there are increased risks of spacesuit arcing.

There are three basic types of particle adhesion mitigation strategies, namely sacrificial coverings, active, and passive systems. Due to the added weight and cost of the sacrificial coating methods, they are a less practical solution. In order to eliminate the buildup of particles, active mitigation systems need external energy (e.g., electrostatic repulsion, sintering into large particles, brush-actuated removal, etc.). Since particle adherence is inherent to the material, passive mitigation solutions don't require external energy. To solve the dust issue, any one of the three types of mitigation strategies may be used alone or in combination.

In nature, the lotus plant, Nelumbo nucifera, receives the most attention when it comes to reducing a material's surface energy for particle adherence mitigation. Water droplets in this situation have very high contact angles, quickly roll off the leaves, and also pick up surface impurities present on leaf during rolling. This combination of hierarchical surface topographies and low surface energy is mimicked in chemical coatings for lunar dust adhesion mitigation. The phenomenon is termed as "Lotus Effect," and has been reproduced by several researchers. In order to precisely change the surface topography of a range of materials, including polymers, composites, ceramics, and metals, available surface area may be reduced using photolithography and laser ablation patterning.

Another example of biomimetics for dust adhesion mitigation comes from the geckos. The geckos have the ability to climb vertical surfaces, even in dusty environments, owing to the hierarchical surface topography of their toes. The particles trapped in their toes are removed by simply stepping on a clean surface. Since contact self-cleaning is to be accomplished in dry conditions, the production of biomimetic materials that mimic these qualities is especially attractive for reducing lunar dust adhesion. The gecko-mimetic polypropylene surfaces contaminated with silica-alumina microspheres (size ~ 1–5 μm) were studied using a similar approach.

The efficiency of surface modified materials to reduce particulate adhesion was evaluated using two methods, each concentrating on either single-particle interactions or multiple-particle interactions. Atomic force microscopy (AFM) was used to measure the force of adhesion between single particles, while a customized equipment was used to measure multiple-particle adhesion. The particles adhered to the surface were separated using the sonication process in this device [7].

2.7 *Atomic Oxygen Resistant Coatings for LEO*

The atomic oxygen (AO) degrades the polymers and composites used for satellites in the LEO environment. Therefore, these polymers and composites need protection using AO-resistant layer. Mechanical damage to the AO-resistant coatings, however, can make the underlying polymers susceptible to erosion and reduce their service life. Therefore, self-healing AO-resistant coatings are being explored that have the capability to healing when some mechanical damage is done under LEO environment. The UPy-POSS polymeric coatings have a strong adhesion to polymers and are mechanically durable, thermally stable, and transparent. The formation of the epidermal SiO_2 layer following AO exposure is the reason why these self-healing coatings have exceptional resistance against AO [36]. After the silica coating is applied superficially, the erosion yield of Kapton is reduced by nearly three orders of magnitude. Also the surface of coating is uniform and smooth, after AO exposure and no surface shrinkage induced cracks or erosion are observed [37].

Flexible protective coatings for solar panels, silver interconnects, and fiberglass structural components have been identified as a critical technological gap. A number of protective carborane and siloxane coating systems have been developed and evaluated to bridge this gap. It is possible for each carborane unit to incorporate up to 15 oxygen atoms, owing to its unique structure. Contrary to siloxane, which breaks when exposed to AO, the carborane (siloxane) coating showed no signs of cracking. Surface analysis of coatings exposed to the LEO environment reveals the formation of a glassy borosilicate layer, providing excellent protection to substrate [38].

The life expectancy of current generation satellites is limited to 3–5 years, owing to the environmental degradation of the structural materials. The reactive atomic oxygen, along with manmade debris and natural micrometeoroids, large temperature extremes, and ultraviolet radiation have a synergistic degradation effect on most of the polymers and their composites, causing rapid degradation.

3 Zero Gravity Research Facility

Since its inception in 1966, the Zero-G Research Center has served as NASA's top ground-based microgravity research facility. It was initially built during the 1960s space race to facilitate research and development of fluid systems and spaceflight components in a weightless or microgravity environment. NASA-funded researchers from around the world are currently using the facility to develop and test experiment hardware intended for flight aboard the International Space Station or future spacecraft. They also investigate the effects of microgravity on physical phenomena like combustion and fluid physics and develop and demonstrate new technology for future space missions [39].

4 Future Trends

A remarkable nanocarbon called graphene has a few layers of carbon atoms arranged in six-membered rings. The mono- and bi-layer graphene exhibits extraordinarily unique electronic properties. It is also important to highlight that graphene has other features, including those related to gas adsorption, magnetic and electrochemical characteristics, and the effects of doping with electrons and holes [40]. These remarkable properties intersect with unique aerospace requirements, rendering it an exciting candidate for such applications.

But before they can be used in space constructions, graphene-based structures must overcome a number of obstacles. The absence of a large-scale technique for consistently and repeatedly producing high-quality graphene is one such obstacle. Before a novel material can be taken into consideration for use in space vehicles, it must undergo considerable development, testing, qualification, and certification.

This is usually done with material readily available in large quantities with minimal variability in batch-to-batch properties [41].

In addition to improved performance and reduced weight, graphene has the potential to meet some critical needs in the aerospace systems. The atomic-layer thickness, mechanical strength, and high electron mobility in graphene advocate its suitability for flexible electronics. These devices can be easily integrated into multifunctional structures. The mono-layer atomic thickness of graphene offers additional benefit of intrinsic radiation hardness, which can drastically reduce the amount of conventional radiation shielding material. Many aspects of graphene are yet to be explored, including its applications for zero-G applications. In this latest experiment, the researchers tried to use graphene's excellent thermal properties to improve the performance of cooling systems in use in satellites [42].

References

1. May, S.: What is a satellite? | NASA [Internet]. Available from: https://www.nasa.gov/audience/forstudents/5-8/features/nasa-knows/what-is-a-satellite-58.html, cited 15 Apr 2022
2. Applications of seals and polymer material solutions in satellite space systems. AZO Mater [Internet]. https://www.azom.com/article.aspx?ArticleID=19341, 2 Jul 2020
3. Perez, R.: Introduction to satellite systems and personal wireless communications. In: Perez, R. (ed.) Wireless Communications Design Handbook, pp. 1–30. Academic Press (1998)
4. Small satellite market size to reach USD 13711.7 million by 2030 at CAGR 16.4%—valuates reports [Internet]. https://www.prnewswire.com/in/news-releases/small-satellite-market-size-to-reach-usd-13711-7-million-by-2030-at-cagr-16-4-valuates-reports-803196164.html, cited 15 Apr 2022
5. Grossman, E., Gouzman, I.: Space environment effects on polymers in low earth orbit. Nucl. Instrum. Methods Phys. Res. Sect. B Beam Interact. Mater Atoms **208**, 48–57 (2003)
6. Krishnamurthy, V.N.: Polymers in space environments. Polym. Other Adv. Mater. 221–226 (1995)
7. Wohl, C.J., Connell, J.W., Lin, Y., et al.: Generation and evaluation of lunar dust adhesion mitigating materials. In: 3rd AIAA Atmos Sp Environ Conf., pp 1–12 (2011)
8. Willis, P.B., Hsieh, C.H.: Space applications of polymeric materials. Kobunshi. **49**, 52–56 (2000)
9. Naser, M.Z.: Extraterrestrial construction materials. Prog. Mater. Sci. **105** (2019)
10. Naser, M.Z., Chehab, A.I.: Polymers in space exploration and commercialization [Internet]. Polym. Sci. Innov. Appl. INC (2020). https://doi.org/10.1016/B978-0-12-816808-0.00014-7
11. Dagras, S., Eck, J., Tonon, C., et al.: Adhesives in space environment. In: Handbook Adhesive Technology, vol. 2–2, 2nd edn, pp. 915–940 (2018)
12. Hager, M.D., Bode, S., Weber, C., et al.: Shape memory polymers: past, present and future developments. Prog. Polym. Sci. **49–50**, 3–33 (2015)
13. Bram, A.I., Gouzman, I., Bolker, A., et al.: The effect of POSS type on the shape memory properties of epoxy-based nanocomposites. Molecules **25** (2020)
14. Herath, H.M.C.M., Epaarachchi, J.A., Islam, M.M., et al.: Carbon fibre reinforced shape memory polymer composites for deployable space habitats. Eng. J. Inst. Eng. Sri Lanka **52**, 1 (2019)
15. Bheekhun, N., Rahim, A., Talib, A., et al.: Aerogels in aerospace: a review. Adv. Mater. Sci. Eng. **2013**, 1–18 (2013)
16. Shaker, K., Jabbar, A., Karahan, M., et al.: Study of dynamic compressive behaviour of aramid and ultrahigh molecular weight polyethylene composites using Split Hopkinson Pressure Bar.

J. Compos. Mater. [Internet] **51**, 81–94 (2022). http://jcm.sagepub.com/cgi/doi/10.1177/0021998316635241
17. Mangalgiri, P.D.: Composite materials for aerospace applications. Bull. Mater. Sci. **22**, 657–664 (1999)
18. Nawab, Y., Sapuan, S., Shaker, K. (eds.): Composite Solutions for Ballistics, 1st edn. Elsevier, Cambridge, USA (2021)
19. Rawal, S.P., Misra, M.S., Wendt, R.G.: Composite Materials Applications for Space. Hampton, Virginia (1990)
20. Rehman, M.M., Shaker, K., Nawab, Y.: Effect of PEEK particles on physiomechanical behavior of carbon/epoxy composite. Int. J. Polym. Sci. **2022**, 12 (2022)
21. SAMPE: Composite applications for space [Internet]. https://www.nasampe.org/page/CompositesApplicationsforSpace, cited 9 Apr 2022, p. 365 (2019)
22. Rivera Lopez, M.Y., Lambas, J.M., Stacey, J.P., et al.: Development of cycloaliphatic epoxy-POSS nanocomposite matrices with enhanced resistance to atomic oxygen. Molecules **25** (2020)
23. Winroth, S., Scott, C., Ishida, H.: Structure and performance of benzoxazine composites for space radiation shielding. Molecules **25** (2020)
24. Shaker, K., Nawab, Y.: Lignocellulosic Fibers Sustainable Biomaterials for Green Composites. Springer, Cham, Switzerland (2022)
25. Sreekala, M.S., STPBMJJLPo.: Natural fibre composites and their applications in aerospace engineering. In: Rana, S., Fangueiro, R. (eds.) Advanced Composite Materials for Aerospace Engineering, pp. 365–383. Woodhead Publishing (2014)
26. Campbell, D., Lake, M.S., Scherbarth, M.R., et al.: Elastic memory composite material: an enabling technology for future furlable space structures. In: Collect Tech Pap—AIAA/ASME/ASCE/AHS/ASC Struct Struct Dyn Mater Conf. vol. 10, pp. 6735–6743 (2005)
27. Liu, Y., Du, H., Liu, L., et al.: Shape memory polymers and their composites in aerospace applications: a review. Smart Mater. Struct. **23** (2014)
28. Toor, Z.S.: Applications of composite materials. J. Sp. Technol. **8**, 65–70 (2018)
29. Kubit, A., Trzepiecinski, T., Krasowski, B., et al.: Strength analysis of a rib-stiffened GLARE-based thin-walled structure. Materials (Basel). 13 (2020)
30. Salve, A., Kulkarni, R., Mache, A.: A review: fiber metal laminates (FML's)—manufacturing, test methods and numerical modeling. Int. J. Eng. Technol. Sci. **3**, 71–84 (2016)
31. Hussain, M., Imad, A., Nawab, Y., et al.: Effect of matrix and hybrid reinforcement on fibre metal laminates under low–velocity impact loading. Compos. Struct. (2022)
32. Patil, N.A., Mulik, S.S., Wangikar, K.S., et al.: Characterization of glass laminate aluminium reinforced epoxy—a review. Proc. Manuf. [Internet] **20**, 554–562 (2018). https://doi.org/10.1016/j.promfg.2018.02.083
33. Bieniaś, J., Jakubczak, P.: Impact damage growth in carbon fibre aluminium laminates. Compos. Struct. [Internet] **172**, 147–154 (2017). https://doi.org/10.1016/j.compstruct.2017.03.075
34. Boudjemai, A., Amri, R., Mankour, A., et al.: Modal analysis and testing of hexagonal honeycomb plates used for satellite structural design. Mater. Des. [Internet] **35**, 266–275 (2012). https://doi.org/10.1016/j.matdes.2011.09.012
35. Ion, D., Adriana, S., Ana, S.: Aluminum/glass fibre and aluminum/carbon fibre hybrid laminates. Incas Bull. **2**, 33–39 (2010)
36. Wang, X., Li, Y., Qian, Y., et al.: Mechanically robust atomic oxygen-resistant coatings capable of autonomously healing damage in low earth orbit space environment. Adv. Mater. **30**, 1–7 (2018)
37. Hu, L., Li, M., Xu, C., et al.: Perhydropolysilazane derived silica coating protecting Kapton from atomic oxygen attack. Thin Solid Films **520**, 1063–1068 (2011)
38. Schwam, D., Litt, M.H.: Evaluation of atomic oxygen resistant coatings for space structures. Adv. Perform. Mater. **3**, 153–169 (1996)
39. Zero Gravity Research Facility [Internet]. https://www1.grc.nasa.gov/facilities/zero-g/, cited 9 Apr 2022

40. Rao, C.N.R., Biswas, K., Subrahmanyam, K.S., et al.: Graphene, the new nanocarbon. J. Mater. Chem. **19**, 2457–2469 (2009)
41. Siochi, E.J.: Graphene in the sky and beyond. Nat. Nanotechnol. [Internet]. **9**, 745–747 (2014). https://doi.org/10.1038/nnano.2014.231
42. Zero gravity graphene: testing space applications | Shaping Europe's digital future. https://digital-strategy.ec.europa.eu/en/news/zero-gravity-graphene-testing-space-applications. 2017 Jul 12, cited 9 Apr 2022

Modern Characterization Techniques for Functional Polymers

H. M. Fayzan Shakir and Rukhsar Anum

Abstract Polymers and their composites are preferred over conventional materials like steel, copper, and aluminum due to their high corrosion resistance, flexibility, ease of processability, and lightweight. Various analytical techniques have been used for the characterization of polymers and their composites as a function of either time or temperature. Morphological structure, molecular weight, analysis of monomer, solvent residue, the composition of the copolymer, and interfacial interfaces of polymeric systems are some of the advanced properties of polymers and composites. The current chapter covers some of the advanced characterizations techniques like Dynamic Mechanical Analysis (DMA), Thermomechanical Analysis (TMA), Atomic Force Microscopy (AFM), 4-Probe technique, Inverse Gas Chromatography (IGC), and Gel Permeation Chromatography (GPC). The above-mentioned techniques are used to determine various properties of advanced polymers like their response to heat, stress, electrical resistance, molecular weight, etc.

1 Introduction

An extensive range of analytical techniques has been established for polymer characterization and their composites across the time. Most of these techniques have been used for identifying the physical properties while some of these are used for determining the chemical structure. The polymers are characterized by these different techniques as either function of time or temperature. Thermo-analytical techniques are generally used for the identification of physical properties. Thermal analysis is an enormously diverse technique with a wide range of different methods. Thermomechanical Analysis (TMA) is a sensitive and ideal thermal analytical technique that provides information about thermal expansion, softening points, and phase changes

H. M. Fayzan Shakir (✉)
Department of Materials, School of Engineering and Technology, National Textile University, Faisalabad, Pakistan
e-mail: Fayzan.shakir@ntu.edu.pk

R. Anum
Department of Chemistry, University of Agriculture Faisalabad, Faisalabad, Pakistan

K. Shaker and A. Hafeez (eds.), *Advanced Functional Polymers*, Engineering Materials,
https://doi.org/10.1007/978-981-99-0787-8_10

in polymers as a function of temperature. Dynamic Mechanical Analysis (DMA) is an analytical technique that provides insight into its chain structure and is used to characterize the interfacial interfaces of polymeric systems.

In recent years, the technique of chromatography is attaining fame in the characterization of polymers. Different techniques such as size exclusion chromatography, gel permeation chromatography, and high-performance chromatography are integral parts of polymer characterization. These are used for the determination of molecular weight, analysis of monomer, solvent residue, composition of the copolymer, etc. Another technique gaining importance in polymer analysis is inverse gas chromatography for the investigation of residual monomers and impurities in the polymer [1]. Some of the characterization techniques for the polymer are explained for better understanding.

2 Dynamic Mechanical Analysis

Dynamic mechanical analysis (DMA) is an analytical practice that provides insight into chain structure and inter-phase interaction in polymer blends. For complex materials like advanced polymer composites, during the DMA spectra interpretation, some attention is essential, especially while determining the glass transition temperature (T_g). For measuring the dynamic mechanical properties in filler-filled polymers, DMA could be considered as one of the standard equipment.

In the DMA experiment, at a given frequency and temperature, an oscillating force is functional to the test material while the response of the material to the applied force is measured [2]. Stress (σ) is the applied force on the sample and strain (γ) is the deformation in the sample. For viscoelastic samples like polymers, the response of material magnitude (i.e., deformation amplitude) to the oscillating force applied is moved through phase angle (δ) shown in Fig. 1. The tan δ, loss, and storage modulus for a free coating sample analyzed by DMA are shown in Fig. 2. The relation between strain and applied stress is created in the material, both the loss modulus (E'') as well as elastic storage modulus (E') are calculated. E'' exhibits the ability of sample to dissipate or lose energy and E' exhibits the capability of the test material to return or store energy. The relation among both factors is called damping and expressed as $\tan\delta\left(\delta = \frac{E''}{E'}\right)$.

Different techniques are used to deform the samples in DMA e.g., shear stress, tension, or parallel plate blending. The peak values of tan δ are utilized to exhibit the inner friction of the polymeric chain quantitatively. With particular particle size, surface area, and filler's loading, the number of mobile chain segments becomes higher with the outstanding dispersion of filler material in the matrix than in a system with poor dispersion of the filler. Increased mobile chain segments enhanced the inner friction and loss modulus, giving a high peak value at T_g of tan δ. When the concentration of particles in the polymer increased, the peak value of tan δ reduced because of an increase in interfacial interaction [3]. However, DMA results provide

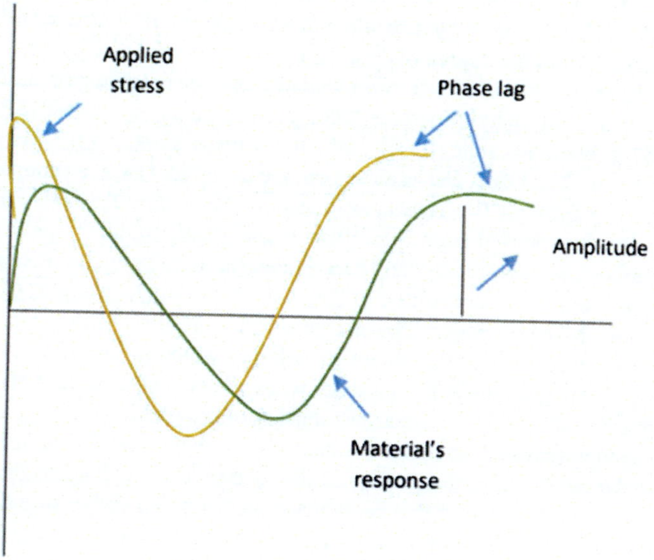

Fig. 1 Schematic illustration for applied stress to sample and response of sample in DMA

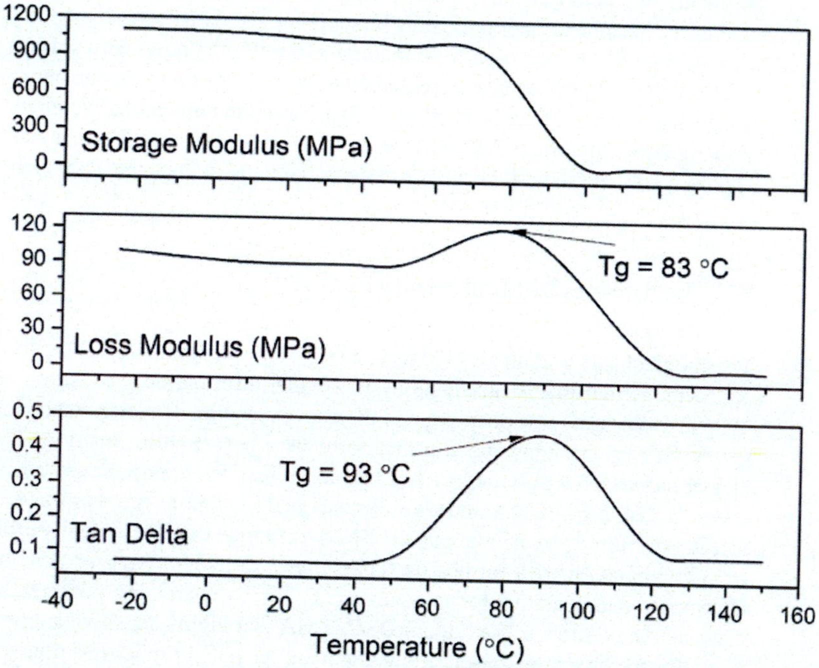

Fig. 2 Tan δ, loss and storage modulus results of a sample analyzed in DMA

useful information related to the extent of interfacial interaction in the region of interphase and the degree of dispersion of the fillers.

Eisenberg et al. [4] showed that the nanocomposites of silica with various polymers poly(methyl methacrylate), styrene-butadiene rubber (SBR), polystyrene (PS), polyvinyl acetate (PVA), and poly(dimethylsiloxane) (PDMS) could have two distinctive T_{gs}. DMA was used to examine the effect of silica nanoparticle content on several polymers. The T_g was estimated by the peak of the tan δ. The concentration of silica was varied from 0 to 50 wt% and all the filled composites exhibited two separate Tgs. The researchers recognized the first Tg as the majority of polymers and the second T_g accredited to the chain of polymers positioned in the interphase section of the composite. The interphase section consists of two nanolayers, the first nanolayer consists of the chains of polymer which are fixed due to the connection with the charged surface of particles while the next layer is thicker and formed away from the nanoparticle surface and consists of polymeric chains which are less bound to nanoparticles. With increased nanosilica content, the T_g location did not change but the tan δ value decreased and this decrease is credited to the decrease in the polymer chain fraction that participates in the major conversion through increased silica amount in the composite. The second peak of tan δ was wider and depending on the type of polymer was located about 40–110 °C high than that of the first one. The second T_g and second peak were reduced due to increased silica loading. The area below the two peaks of tan δ reduced with silica content, i.e., at high filler content, tan δ curve became thinner indicating lesser polymer chains participated in Tg. It was also suggested that with high filler loading >20 wt%, the polymer's volume fraction bound with the filler surface increased causing a reduction in the tan δ curve width. It was also exhibited that the second Tg depends on the molecular weight and nature of the polymer and the nanofiller loading in the composite [5]. Further scientists [6, 7] presented that the major or inmost interfacial layer defines the Tg for composites.

3 Thermomechanical Analysis

Thermomechanical analysis (TMA) is a reliable, excellent, simple, sensitive, basic, and ideal thermal analytical technique. It provides information about softening points, T_g, phase changes, composition, and thermal expansion of materials having several geometries by applying the constant force as a temperature function. TMA is an elementary tool for polymer science because it can detect proficiently and precisely every probable transition and also dimensional change at the nano level [8]. The investigation study explained that the TMA technique can be efficiently utilized only or in combination with further TA techniques like DMA, TGA, or DSC because it is accurate and sensitive for recognizing T_g for polymers and polymer composite materials in contrast to traditional DSC. TMA is a highly sensitive analytical technique for the characterization of chemical and physical properties of polymers and composite materials that are unnoticeable [9]. The purpose of T_g from TMA, DMA, and DSC procedures are comparatively shown in Fig. 3. Extraordinarily, T_g for the

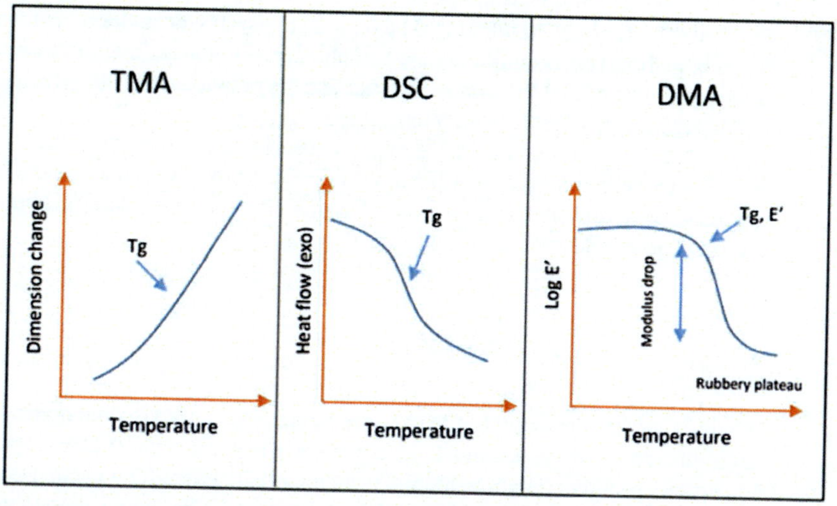

Fig. 3 Comparison of T_g from TMA, DSC, and DMA techniques

sample could be attained from the CTE curve, i.e., TMA graph by detecting the two curves of intercept plot among displacement versus temperature shown in Fig. 4.

TMA plot is efficiently utilized to explore T_g for polymers or polymeric nanocomposites as a polymer changes from a glassy state to rubbery through a modification in permitted molecular volume regarding CTE alteration as a function of pressure at temperature constant or a function of temperature at constant pressure [10]. The thermo mechanical characteristics of basalt fiber (BF) filled high-density polyethylene (HDPE) and co-extruded wood plastic composite with the composite shell of HDPE/BF were investigated. It was reported that the introduction of BF in HDPE

Fig. 4 T_g plot from TMA technique

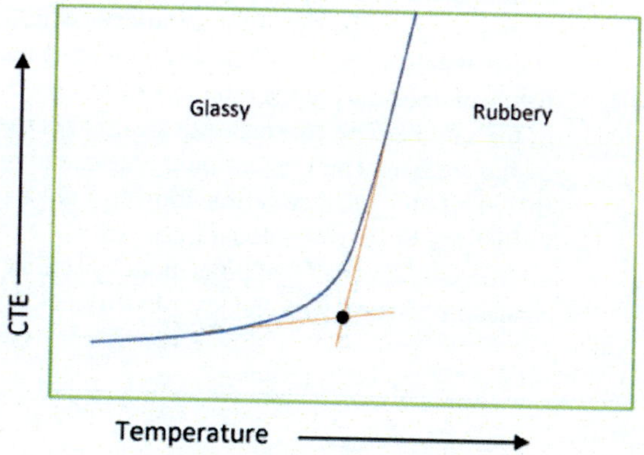

reduced CTE for pure HDPE [11]. The thermal expansion behavior of nanocomposites was analyzed by the introduction of graphene sheets and graphene foam into polydimethylsiloxane (PDMS). It was detected that the graphene sheet/PDMS composite exhibited higher CTE than the composite of graphene foam/PDMS. Researchers observed it as a promising and attractive material for use in photonic and electronic devices for thermal management [12]. Currently , the TMA technique offer applications in different areas such as electroplating, construction, food packaging, and aerospace industry [13].

4 Atomic Force Microscopy

Atomic Force Microscopy (AFM) has been extensively used to do inquiries into the crystallization of polymers due to its remarkable strength which makes it preferable for advanced studies. High-resolution technology allows the fundamental dimension scale of the lamellar polymeric crystal to be measured. The sample preparation technique is simple as it demands no metal coating and staining. AFM is non-destructive in several conditions. The images can be obtained during the melting and crystal growth that provide lamellar and sub-lamellar resolution with respect to time. AFM provides polymer crystallization with crystal melting, crystal growth, and crystal organization at the lamellar scale that represents structure progress and influence of kinetics at local conditions. It has several measuring modes with the addition of various functional semi-crystalline polymers [14].

A sharp tip is present in AFM which is attached to the cantilever beam that makes contact with the surface of the sample. There are several modes of the technique but 'contact mode' as well as 'tapping mode' are frequently used. The measurement of the force of attraction by deflection of the cantilever between the tip and surface is done in contact mode which is generally monitored with an optical lever. The deflection of the cantilever is adjusted by the height of the base through the feedback loop. The force acting on the tip of the cantilever is known as linear spring at set point. The constant deflection is sustained by regulating the location of the cantilever through the feedback loop during the scanning of the cantilever point by point above the surface of the sample. The control vertical signal is used to synthesize a topographic image of the surface. The contact mode maintaining the interface of the tip and sample that exerts the adjacent attractions on the sample with tip is quite great, generally destroying loosely or soft samples [15].

The cantilever of 'tapping mode' oscillate at its resonance frequency. The tip makes interaction with the sample surface damped the amplitude of the oscillation with the contact. The amplitude of oscillation is a constant value and is used as a feedback factor. The stable image of a soft sample of polymer can be obtained under the careful condition where the lateral forces are lesser than the contact mode of AFM. A set frequency driving the cantilever and contact with the properties of the material surface could be attained by observing the variance between the driven signal phase and the cantilever response phase—the phase image [16].

The excellent mode of technique is used to analyze the polymer crystal growth that has very sharp variations in the mechanical and adhesive characteristics among liquid and crystalline polymers. The monitoring of sample temperature is very necessary for the melting and crystallization of the polymers. The process of the stable and constant temperature of AFM has allowed an extensive range of polymers to attain images at the temperature of 0–250 °C during crystallization. The polymers susceptible to hydrolysis and oxidation are studied at high temperatures by use of an inert gas atmosphere and vacuum for environmental control. The crystallization of many polymers in AFM is allowed at the temperature range near T_g or Tm when the growth of the crystal is low [17] (Fig. 5).

Recently, an advanced form of AFM is developed that permits images to be composed at video rate and is known as Video-AFM. This advanced technology helps in understanding some progress for the growth of crystal but is troubled due to large tip-sample forces because Video-AFM needs constant tip-sample contact. The rate of at least 1 frame is needed for the recurrent contact mode image in the future advancement of this technique because fast scanning helps to understand the significant impact on our understanding of polymeric crystal growth in various varieties of systems.

Atomic force microscopy is an old technique and is advancing constantly with high-speed scanning, high-resolution imaging, and process of material mapping that are potentially important for the future applications of AFM to polymer crystal [18] (Fig. 6).

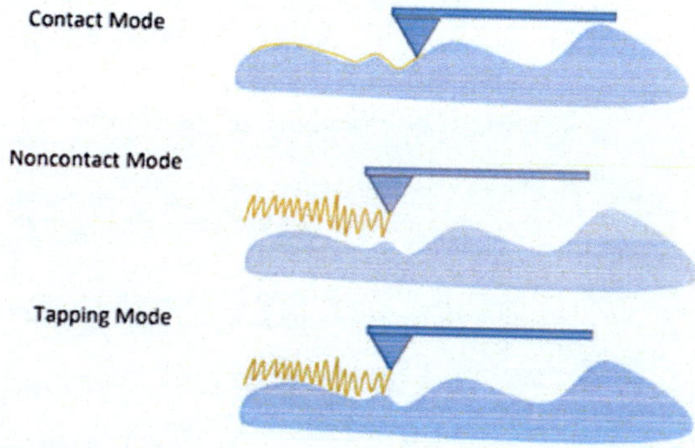

Fig. 5 Different modes of AFM

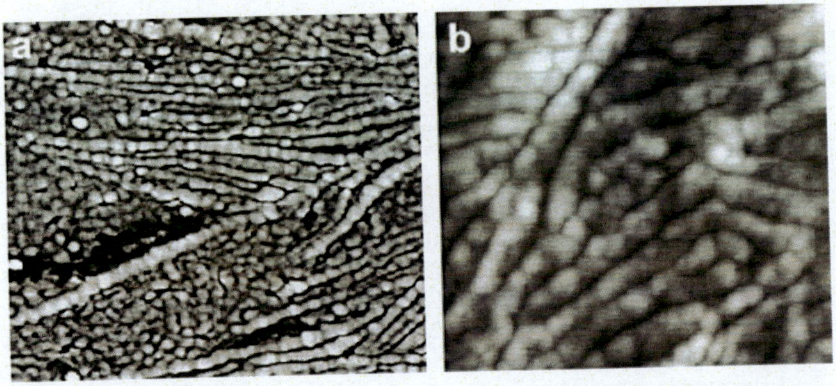

Fig. 6 AFM phase images of semi-crystalline polymer surface representing the observed 'knobbly' substructure that has been thought due to its early crystallization in discrete blocks. Scale bars show 200 nm in **a** and 100 nm in **b** [19].

5 Probe Technique

The 4-probe technique is used to measure the conductivity of polymer materials. It can measure the conductivity of the material in either film or bulk. The instrument consists of equally sized tungsten metal tips with a fixed radius. Every strip is sustained by a spring on the other edge to reduce the damage to testing material during probing. All the metal tips are part of the auto-mechanical step which moves up and down during the process. The outer two probes provide the current by high impedance current source while the inner two probes determine the voltage through a voltmeter for measuring the resistivity of the sample polymer (Fig. 7).

The resistance is directly proportional to the length and inversely proportional to the area of the cross section of testing material at a constant temperature.

$$R = R = \rho L/A \tag{1}$$

ρ represents the resistivity of the conducting material and is measured in an ohm-meter.

$$\sigma = 1/\rho \tag{2}$$

Conductivity is reciprocal to resistivity and its temperature dependence is given by

$$\rho = A \exp \frac{E_g}{2KT} \tag{3}$$

E_g represents the band gap of the material, T represents the temperature in kelvin, and K represents the Boltzmann constant (8.6×10^{-5} eV/K).

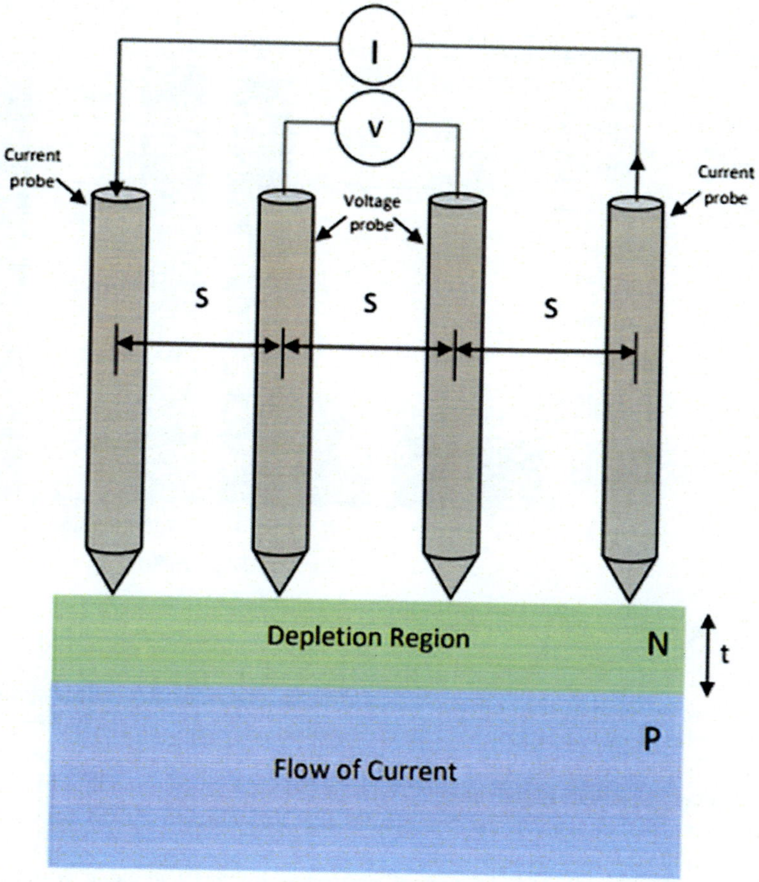

Fig. 7 4-probe technique

Yamanaka et al. measured the electrical conductivity of 3D polymer with dimensions of $0.2 \times 0.2\ mm^3$ at temperature range of 4–300 K by 4-probe technique [20]. Vas and Thomas also measured the electrical conductance of polymeric composite with increasing nanofiller of MWCNT by using 4-probe technique as shown in Fig. 8. Agilent device analyzer B1500A was used for these measurements.

6 Inverse Gas Chromatography

IGC is an accurate, fast, and reliable system for the physiochemical characterization of polymers, and their composites. The word 'inverse' represents the sample material

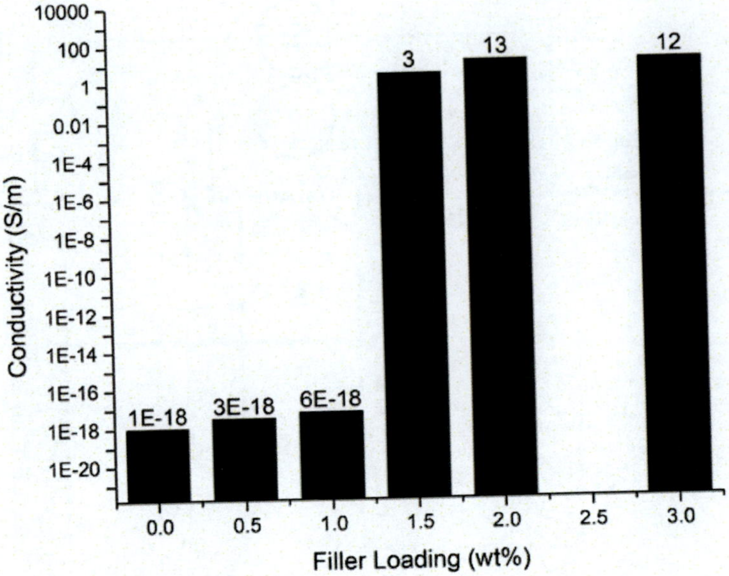

Fig. 8 Conductivity measurements for polymeric composite with nanofiller loading [21]

such as a polymer composite is a material of interest existing in the chromatographic column, in comparison to general analytical gas chromatographic experiments. The sample solute is carefully selected and injected into the tube of the carrier gas flowing over the surface of the sample. The peak elution and retention time of selected solutes are used to analyze the interactions that are affected by the contacts among the stationary phase and solutes.

The investigation of surface properties by this technique is quite interesting. Surface properties are commonly characterized by the use of factors that describe the ability of the surface to experience dispersive and acid-base characteristics. Free energy of the surface has a dispersive component of γ_s^d and specific interaction factors counting the ability of the surface to behave as a donor or acceptor are utilized for measuring the characteristics of polymeric surfaces [22] (Fig. 9).

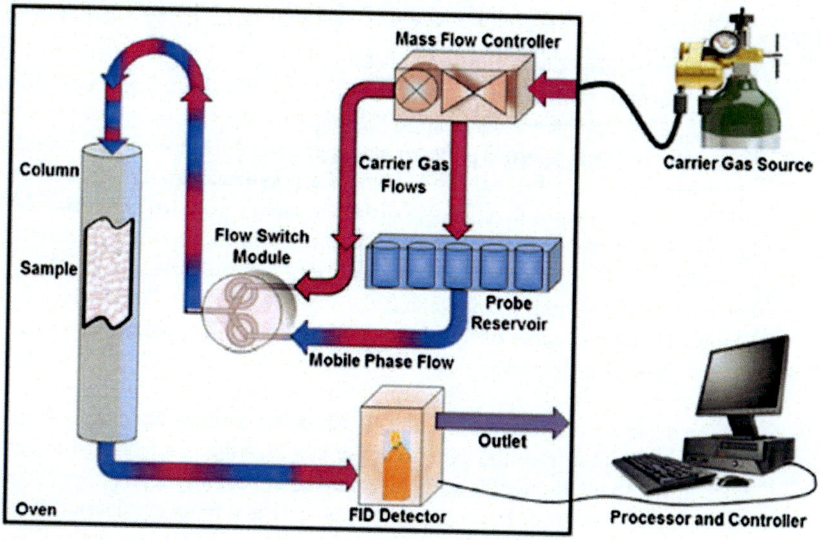

Fig. 9 Schematic representation of inverse gas chromatography analysis[23]

The dispersive component can be measured from Eq. (4) with experimentally determined value of ΔG_{CH_2}.

$$\gamma_s^d = \frac{\Delta G_{CH_2}}{4(6.023 \times 10^{23})^2 a_{CH_2}^2 \gamma_{CH_2}} \tag{4}$$

While

$$\Delta G_{CH_2} = -RTln\left(\frac{V_{N,n}}{V_{N,n+1}}\right) \tag{5}$$

$V_{N,n}$ and $V_{N,n+1}$ represents the net volume retentions of alkanes having n and $n + 1$ atoms respectively, a_{CH_2} represents the surface covered by polymer sample and γ_{CH_2} represents the free surface energy of the polymer.

Schultz et al. proposed the calculation of γ_s^d from the following equation:

$$RTlnV_N = 2N(\gamma_s^d)^{1/2} a_p (\gamma_L^d)^{1/2} + c \tag{6}$$

a_p represents the occupied area by the adsorbing molecule of test material, γ_L^d is the dispersive component of adsorbent energy of the free surface, and c is constant. The technique is based on the linear relationship of ΔG^o and $a_p(\gamma_L^d)^{1/2}$ of alkanes and determination of γ_s^d from slop value.

Various restrictions for this technique have been indicated by Brendle and Papirer. The problems arose due to the proper determination of $a_p\left(\gamma_L^d\right)^{1/2}$ value, i.e., accurate measurement of inverse gas chromatography. The value of γ_s^d may be influenced by the different parameters such as applying reference state in the process and impact of bulk retention during porous polymer having temperature near to T_g or when a sample solute is utilized as solubilizer and solvent for polymeric sample. For reducing these problems, it must guide that the porosity of the sample polymer should be checked for its influence on the retention factor of sample solute and exertion at the temperature distance. These errors depend on the quality of determination of the slope value in the finding of dispersive components for the free energy of the surface in Eq. (6). Generally, the low value of the slope is multiplied by a constant value greater than three orders. The 10% error in the slop value results in the 20% inaccuracy in the γ_s^d values.

The experimental value of V_g and T can be determined accurately. The value of $a\gamma_L^d$ can be found from the experiment or can be collected from the literature. The main difficulty is finding the value of a_p that can be a larger error depending on the temperature, properties of the adsorbent as well as reference material. Its value can be obtained with one of these assumptions: (1) the test material may act as an ideal gas; (2) the value of a_p is constant; (3) at a temperature below the boiling point the test material act as real gas; (4) the test material act as a liquid when the adsorption depends on the density of the test material. Different models for the temperature dependence of a_p are present but these assumptions cannot complete the conditions of outcomes. The constant value of a_p is a most suitable assumption but in inverse gas chromatography, the behavior of test material as ideal gas seems to be realistic.

The significance of using IGC can be summarized as follows: (1) these are promising in checking the effect of modified arrangement on estimated factors and additional characteristics of sample materials; (2) this technique allows the examination of various porous polymers that have hardly determined surface activity, e.g., measurement of contact angle; (3) evaluation of the relationship between the surface properties and monolayer; (4) test of the effect of IGC experiment temperature on the surface is possible; (5) the alteration of surface properties may occur as the outcome of surface modification [24].

7 Gel Permeation Chromatography

Gel permeation chromatography (GPC) or size exclusion chromatography (SEC) having a refraction index as well as a viscosity detector is most important to attain information regarding the molecular weight of polymers. This technique is fast and runs more routinely in comparison to 13C NMR. When joined with multidetector allows attaining detailed as well as fast characterization since different involved techniques give structural data which are attained instantaneously with standard GPC analysis. Suarez et al. [25] characterized the synthesized copolymers using the

GPC-4D technique by coupling GPC with four detectors: a viscometer, a refractive index, an infrared, and a multi-angle light scattering (MALS) detector.

The chromatography of gel permeation is Water Alliance 2000 having standard viscosity and refractive index detectors. In GPC, the separation was done by two columns, the flow rate was mLmin^{-1}, and the temperature was fixed at 145 °C. The solvent 1,2,4-trichlorobenzene (TCB) was added for stabilizing the polymer in opposition to oxidative degradation.

MALS detector was equipped at 690 nm with a laser and 17 multi-angle detectors. The temperature was 145 °C and the joining between the MALS cell and GPC was carried out by using the heated line. Toluene was used for the calibration of the DAWN EOS system and the normalization of detectors was done using the polystyrene (PS) standard. Wyatt Technology software was used for the acquisition of online data and calculation of the average and distribution of the radius of gyration and molecular weight.

An infrared detector (IR) is a polymeric char detector joined with GPC by a heated transfer line at a temperature of 145 °C. It gives different wavelengths analysis equivalent to absorption at several IR bands which helps attain the intensity for the characteristic peaks of CH and CH_3.

Homo and copolymers using several catalysts were fabricated covering a 0–50 mol% composition range of ethylene. For metallocene copolymers, with ethylene percentage, the molecular weight was decreased which was because of polymer crystallinity. For the Ziegler-Natta catalyst, the average molecular weight was high, and not a clear trend was seen with the copolymer composition. Classic results attained using the GPC-4D system are shown in Fig. 5. The molecular weight as well as molecular weight distribution (MWD) for every GPC slice is attained through a universal calibration curve by viscosity and refractive index detectors. Comonomer content attained through IR and the radius of gyration for every molecular weight slice attained through MALS are displayed. The comonomer distribution attained for copolymers is like that determined through the IR5-MCT detector [26], which confirmed the GPC-4D technique's accuracy. Therefore, it is promising to obtain broad information regarding comonomer distribution, molecular weight, and copolymer confirmation in solution [27] (Fig. 10).

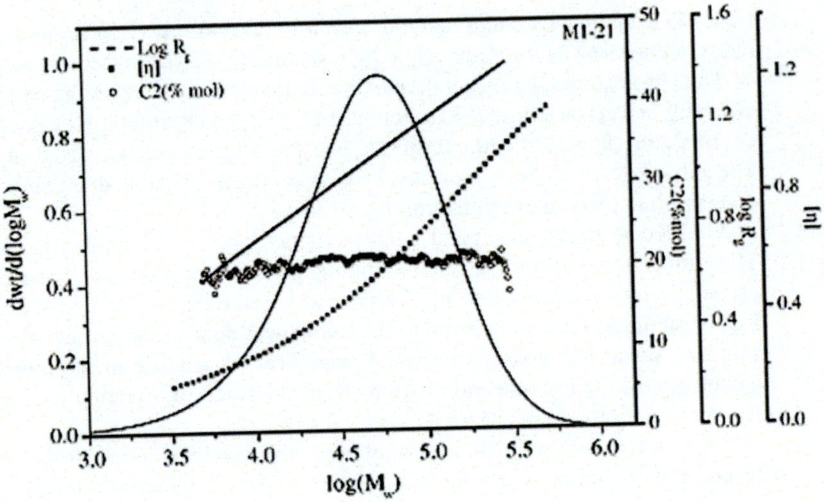

Fig. 10 Distribution of ethylene content, MWD, intrinsic viscosity, and radius of gyration for copolymers by GPC-4D technique

References

1. Van Lieshout, M.H.P.M., Janssen, H.G., Cramers, C.A., Hetem, M.J.J., Schalk, H.J.P.: Characterization of polymers by multi-step thermal desorption/programmed pyrolysis gas chromatography using a high temperature PTV injector. HRC J. High Resolut. Chromatogr. **19**, 193–199 (1996). https://doi.org/10.1002/jhrc.1240190404
2. Liang, J.Z.: Dynamic mechanical properties and characterization of inorganic particulate-filled polymer composites. J. Thermoplast. Compos. Mater. **24**, 207–220 (2011). https://doi.org/10.1177/0892705710387254
3. Bashir, M.A., Jakobsen, M.G., Farstad, V.B.: The effect of extender particle size on the glass transition temperature of model epoxy coatings. Polymers (Basel) **12** (2020). https://doi.org/10.3390/polym12010196
4. Tsagaropoulos, G., Eisenberg, A.: Dynamic mechanical study of the factors affecting the two glass transition behavior of filled polymers. Similarities Differ. Random Lonomers, Macromol. **28**, 6067–6077 (1995). https://doi.org/10.1021/ma00122a011
5. Bashir, M.A.: Use of Dynamic Mechanical Analysis (DMA) for characterizing interfacial interactions in filled polymers. Solids. **2**, 108–120 (2021). https://doi.org/10.3390/solids2010006
6. Mayes, A.M.: Softer at the boundary taking lessons from the book. Nature **4**, 651–652 (2005)
7. Starr, F.W., Schrøder, T.B., Glotzer, S.C.: Effects of a nanoscopic filler on the structure and dynamics of a simulated polymer melt and the relationship to ultrathin films. Phys. Rev. E Stat. Physics, Plasmas, Fluids, Relat. Interdiscip. Top. **64**(5) (2001). https://doi.org/10.1103/PhysRevE.64.021802
8. Gaisford, S., Kett, V., Haines, P.: Principles of thermal analysis and calorimetry. Roy. Soc. Chem. (2019)
9. James, J.: Chapter 7—thermomechanical analysis and its applications. In: Thomas, S., Thomas, R., Zachariah, A.K., Mishra, R.K. (eds.) Thermal and Rheological Measurement Techniques for Nanomaterials Characterization, pp. 159–171. Elsevier (2017). https://doi.org/10.1016/B978-0-323-46139-9.00007-4

10. Corcione, C.E., Frigione, M.: Characterization of nanocomposites by thermal analysis. Materials (Basel). **5**, 2960–2980 (2012). https://doi.org/10.3390/ma5122960
11. Wu, Q., Chi, K., Wu, Y., Lee, S.: Mechanical, thermal expansion, and flammability properties of co-extruded wood polymer composites with basalt fiber reinforced shells. Mater. Des. **60**, 334–342 (2014). https://doi.org/10.1016/j.matdes.2014.04.010
12. Zhao, Y.-H., Wu, Z.-K., Bai, S.-L.: Study on thermal properties of graphene foam/graphene sheets filled polymer composites. Compos. Part A Appl. Sci. Manuf. **72**, 200–206 (2015). https://doi.org/10.1016/j.compositesa.2015.02.011
13. Saba, N., Jawaid, M.: A review on thermomechanical properties of polymers and fibers reinforced polymer composites. J. Ind. Eng. Chem. **67**, 1–11 (2018). https://doi.org/10.1016/j.jiec.2018.06.018
14. Balzano, L., Kukalyekar, N., Rastogi, S., Peters, G.W.M., Chadwick, J.C.: Crystallization and dissolution of flow-induced precursors. Phys. Rev. Lett. **100**, 1–4 (2008). https://doi.org/10.1103/PhysRevLett.100.048302
15. Hsiao, B.S., Yang, L., Somani, R.H., Avila-Orta, C.A., Zhu, L.: Unexpected Shish-Kebab structure in a sheared polyethylene melt. Phys. Rev. Lett. **94**, 1–4 (2005). https://doi.org/10.1103/PhysRevLett.94.117802
16. Klinov, D., Magonov, S.: True molecular resolution in tapping-mode atomic force microscopy with high-resolution probes. Appl. Phys. Lett. **84**, 2697–2699 (2004). https://doi.org/10.1063/1.1697629
17. Yang, X., Loos, J.: Toward high-performance polymer solar cells: the importance of morphology control. Macromolecules **40**, 1353–1362 (2007). https://doi.org/10.1021/ma0618732
18. Garcia, R., Proksch, R.: Nanomechanical mapping of soft matter by bimodal force microscopy. Eur. Polym. J. **49**, 1897–1906 (2013). https://doi.org/10.1016/j.eurpolymj.2013.03.037
19. Hobbs, J.K., Farrance, O.E., Kailas, L.: How atomic force microscopy has contributed to our understanding of polymer crystallization. Polymer (Guildf). **50**, 4281–4292 (2009). https://doi.org/10.1016/j.polymer.2009.06.021
20. Yamanaka, S., Kubo, A., Inumaru, K., Komaguchi, K., Kini, N.S., Inoue, T., Irifune, T.: Electron conductive three-dimensional polymer of cuboidal C60. Phys. Rev. Lett. **96**, 1–4 (2006). https://doi.org/10.1103/PhysRevLett.96.076602
21. Vas, J.V., Thomas, M.J.: Electromagnetic shielding effectiveness of multiwalled carbon nanotube filled silicone rubber. In: INCEMIC 2015—13th International Conference on Electromagnet Interference and Compatibility Proceeding, pp. 55–59 (2017). https://doi.org/10.1109/INCEMIC.2015.8055846
22. Voelkel, A., Grzeskowiak, T.: The use of solubility parameters in characterization of titanate modified silica gel by inverse gas chromatography. Chromatographia **51**, 608–614 (2000). https://doi.org/10.1007/BF02490820
23. Mohammadi-Jam, S., Waters, K.E.: Inverse gas chromatography applications: a review. Adv. Colloid Interface Sci. **212**, 21–44 (2014). https://doi.org/10.1016/j.cis.2014.07.002
24. Voelkel, A.: Inverse gas chromatography in characterization of surface. Chemom. Intell. Lab. Syst. **72**, 205–207 (2004). https://doi.org/10.1016/j.chemolab.2004.01.016
25. Suarez, I., Caballero, M.J., Coto, B.: A fast and reliable procedure to determine the copolymer composition by GPC-IR: application to ethylene/propylene copolymers and comparison with 13C NMR. Polym. Eng. Sci. **51**, 317–322 (2011)
26. Suárez, I., Caballero, M.J., Coto, B.: Composition effects on ethylene/propylene copolymers studied by GPC-MALS and GPC-IR. Eur. Polym. J. **46**, 42–49 (2010). https://doi.org/10.1016/j.eurpolymj.2009.09.005
27. Suárez, I., Caballero, M.J., Coto, B.: Characterization of ethylene/propylene copolymers by means of a GPC-4D technique, Eur. Polym. J. **47**, 171–178 (2011). https://doi.org/10.1016/j.eurpolymj.2010.11.008

Polymer Recycling Techniques

Muhammad Kashif Bangash

Abstract This chapter includes an introduction to polymers and their recycling methods. Recycling is the most suitable alternative to control and avoid the accumulation of plastic waste in the environment. Polymers are compounded for improved performance which makes them difficult to reuse and/or recycle. Polymers can be categorised into two major groups: thermosetting and thermoplastic. The production and use of thermoplastic polymers are much higher in quantity as compared to thermosets globally. Therefore, the recycling techniques for thermoplastics are of major importance. Primary, secondary and tertiary recycling techniques for thermoplastics are discussed in this chapter. Waste management for thermosets is mostly unavoidable and important from an environmental point of view. Thermal, mechanical, chemical waste recycling techniques for thermosets and the inherent recyclability of thermoset polymers are focused in this chapter. The potential applications of recycled thermoplastics and thermosets are also discussed.

Keywords Pollution · Recycling · Thermoplastics · Thermosets · Recycling systems

1 Introduction

Polymers are a very important class of materials that have become an essential part of life. They are all around us in the form of rubbers, plastics, resins, adhesives, etc. As the word polymer describes, they are composed of several smaller molecular units called monomers linked and/or cross-linked to form higher molecular weight structures called macromolecules. Polymerization refers to a chemical process that allows the monomers to join with or without the evolution of by-products.

Sold polymers are generally categorised into elastomers, thermoplastics and thermosetting polymers, based on their physical properties. Elastomers are rubber-like elastic materials. Thermosetting polymers are generally soft or liquid and can be

M. K. Bangash (✉)
Department of Materials, National Textile University, Faisalabad, Pakistan
e-mail: kashifbangash@ntu.edu.pk

K. Shaker and A. Hafeez (eds.), *Advanced Functional Polymers*, Engineering Materials,
https://doi.org/10.1007/978-981-99-0787-8_11

moulded at room temperature and become hard and infusible when heated. Unlike thermoplastics, thermoset polymers once solidified and shaped cannot be reshaped due to the formation of a 3D cross-linking network structure at a molecular level.

The major structural characteristics that determine the behaviour of polymers include [1]:

1. The extent of rigidity of polymer molecules.
2. The degree of attractive forces between the polymer chains.
3. The degree of cross-linking between the polymer chains.
4. The extent of polymer chains to form crystalline domains.

Polymers are tangled long linear chains of molecules. If the intermolecular forces among these chains are weak, they will move past one another when subjected to pressure and will display plastic flow, while for stronger intermolecular forces the plastic flow will be hindered. These polymers can display plastic flow if heated above their Glass Transition Temperature (Tg) and are called thermoplastics. Thermoplastic polymers are solid materials at room temperature and become soft and fluid upon heating. This property allows thermoplastic materials to be cast or moulded.

Crosslink is a chemical bond to connect polymer chains transversely to create a network. Crosslinking limits the relative translational motions of polymer chains and increases the molecular weight [2]. The semi-fluid polymers that become highly cross-linked under heat and/or pressure to form hard, infusible products are thermosetting polymers. Unlike thermoplastics, thermosets do not soften or reshape if heated above Tg due to a 3D network of bonds interlinking the polymer chains.

Based on source, polymers can be of natural, synthetic, or semisynthetic origin. Natural polymers are obtained from plants and animals. Proteins, fats, starch, cellulose and natural rubber are examples of natural polymers. Semisynthetic polymers are extracted from natural sources through chemical processing such as acetate and nitrate polymers are obtained from cellulose. Almost all natural and semisynthetic polymers are biodegradable and are termed biopolymers. Synthetic polymers are manmade polymers and are produced in labs from different sources such as fossil fuels (crude oil, natural gas and coal). Nylon, polyester, styrene, etc. are examples of synthetic polymers. A vast range of synthetic polymers is now available commercially. However, synthetic polymers are mostly non-biodegradable and hence are not considered environmentally friendly [3].

Historically, polymers remained an essential part of human life as cotton, flax, wool, jute, rubber and silk were available in nature since ancient times. For the first time in history, British explorers reported the use of rubber by the Mayan civilisation in Central America in the fifteenth century. Later in the eighteenth century, Charles Goodyear successfully experimented with the vulcanization of rubber using sulphur. The oldest recorded synthetic plastic, bakelite, was developed by Leo Baekeland in 1907. The discovery of the X-ray crystallography technique in the early twentieth century led to the explanation of the structure of natural cellulosic polymers. In 1920, Herman Staudinger, in his classic paper 'Uber Polymerisation' presented the development of modern polymer theory. This development, later, led to the development of several important synthetic polymers including epoxy, nylons, polyvinyl chloride

(PVC), acrylonitrile-butadiene-styrene (ABS), polymethyl methacrylate (PMMA), polycarbonates (PC), etc.

The idea that polymers are mixtures of long-chain molecules connected was not accepted until the work of Staudinger was recognised in the 1930s. The adoption of a definite chemical structure allowed the researchers to understand the behaviour of polymers based on specific monomers. Further, it enabled the scientists to tailor the properties of polymers to a very considerable degree for a specific practical application. In 1974, Wallace Carothers, Paul Flory and Stephanie Kwolek received Nobel Prize for their work on Neoprene, coil configuration and Aramid polymers [4].

Plastics are mainly composed of linear chain molecules of divalent monomers. Occasionally they are engineered with trivalent or tetravalent monomers in the chain for applications that favour the addition of branches to the main chain. They become viscoelastic above their glass transition temperature and are termed thermoplastics as they can be remodelled into useful products. When the multivalent monomer units increase or the branches join to form a single network and can no longer flow above Tg are termed thermosets [2].

Commercial plastics are much more than the combination of basic monomers but are artfully designed compounds for improved performance. They include low molecular weight additives such as stabilisers for heat, light, antioxidants, antistatic, plasticising and fire retardance. Fillers are added to build up the bulk and improve the mechanical, electrical and thermal performance of plastics [5]. Thermodynamically compatible polymers are also blended in some cases to achieve properties for specific applications. The compounding of polymers plays an important role in their recycling.

Owing to the enormous advantages including lower density, processability and easy shaping, plastics became an essential part of life in the current century. The applications of plastics cover almost all the aspects of life such as the packaging for the food industry, electronic products, house insulations, parts of automobiles, aerospace structures and marine vehicles, toys and furniture, medical and defence tools, etc. [6]. However, the other side of the coin is quite different and dark for polymers on account of slow degradation and irresponsible consumer behaviour which has resulted in heavy contamination of the environment with plastic waste and has endangered the ecological equilibrium of the planet.

2 Polymer Pollution

The global production of plastics was reported at around 234 million tonnes (Mt) in 2000, which increased to around 460 Mt in 2019. This increment in global plastic production corresponded to an increase in plastic waste from 156 Mt in 2000 to 353 Mt in 2019. Around 9% of the total plastic waste is currently recycled, 19% is incinerated, 50% goes to sanitary landfills and 22% remains unmanaged and dumped in uncontrolled dumpsites or leaked into the environment [7].

The COVID-19 pandemic combined with the associated lockdown measures affected the global economy significantly. Global plastics use in 2020 reportedly declined by around 2.2% (10 Mt), which is around 4.5% below the pre-COVID projection for 2020 [8]. However, the pandemic led to the surge of single-use plastics, particularly due to the increased use of protective gears leading to extensive littering. With the end of the pandemic, rapid growth in plastic production and its use is projected which will further add to the plastic waste and ultimate environmental concerns.

It was reported that around 22 Mt of plastic waste was added to the environment as waste by 2019. The mismanagement of plastic wastes is the reason for around 88% of plastic leakage and their conversion to microplastics. The remaining 12% of plastic waste enters the environment as microplastics (plastic particles with a diameter < 5 mm) from operations including tyre abrasion on roads, brake wear and pre-dominantly from textile washing. Studies have proved that the presence of microplastics in the food chain and ambient air has increased the exposure of human beings to plastic waste and related hazards [7, 8].

So far around 109 and 30 Mt of plastic waste have been accumulated in rivers and oceans respectively. Around 6.1 Mt of plastic waste was added to rivers and oceans in 2019 alone. This build-up will eventually allow the increased leakage of plastic particles. Moreover, the revival of plastic waste from rivers and oceans is more difficult as the plastic fragments into smaller pieces. The carbon footprint of plastic waste is substantial. The contribution of plastic waste is 3.4% to global greenhouse gas emissions. Plastics generated 1.8 billion tonnes of greenhouse gases in 2019 mainly from their production of fossil fuels [9].

To avoid the undesired plastic waste accumulation in the environment, several techniques and methods are proposed and experimented with in the recent past. Estimates suggest that 80% of global plastic production is comprised of thermoset resins and thermoplastics. Thermoplastics are generally easy to recycle and can be reused. On the contrary, thermosets are difficult to recycle due to their network structure.

3 Systems for Polymer Recycling

Plastics are macromolecular organic compounds or polymers and can be recycled in the following two basic ways:

1. By splitting them into low molecular weight compounds.
2. Without breaking down the molecular structure.

The environmental degradation of plastics is generally a time-taking process. The four basic mechanisms involved in the degradation of plastics are reported in Table 1 [10]

Table 1 Four basic mechanisms involved in the degradation of plastics

Mechanisms for environmental degradation of plastics	Photodegradation
	Thermo-oxidative degradation
	Hydrolytic-degradation
	Biodegradation by microorganisms

The photo-degradation of polymeric chains begins when the plastics are exposed to sunlight. The UV rays in the sunlight provide the activation energy for the incorporation of oxygen molecules which leads to the thermo-oxidative degradation of polymers. Plastics, then become brittle and fracture into smaller pieces until the polymer chains are small enough to be metabolised by microorganisms. The microorganisms convert the smaller polymer chains into carbon dioxide (CO_2). Environmental degradation of plastics is a very slow process, and it takes at least 50 years to complete the cycle [10].

Incineration is the most common approach to generating energy from plastic waste. The problems associated with incineration are the production of toxic gases and the residues containing lead and cadmium. The environmental degradation of plastics is time-taking while incineration adds to the greenhouse gases therefore both are not very much appreciated. As discussed earlier, the life of plastic improves with compounding which, on the other hand, makes it further difficult to degrade, therefore, alternative methods such as recycling are important [11]. There are several alternative recycling techniques proposed to handle plastic waste with advantages including the elimination of environmental problems and recovery of monomers and energy.

4 Recycling of Thermoplastics

The plastic waste problem is more associated with thermoplastic polymers as most of the plastics globally produced and used are thermoplastics. The number of thermoset polymers used in different industries is also considerable, however, they require a different set of recycling techniques due to their cross-linked structures [12].

Thermoplastics can be categorised as commodity plastics, engineering plastics and high-performance plastics based on the speciality of performance, consumption and cost. Commodity plastics are the most prevalent plastics and are produced in high quantities for applications that do not require specially engineered properties. Most of the world's plastics belong to commodity plastics and are made from the following six polymers: Isotactic Polypropylene (i-PP), Low-Density Polyethylene (LDPE), Polyvinylchloride (PVC), Polystyrene (PS), Polyester, Polyethene terephthalate (PET) and Polyurethane (PU).

Engineered thermoplastics are blended to display properties that are not commonly found in commodity plastics. Acetal (POM), Acrylonitrile butadiene styrene (ABS), Polyether ether ketone (PEEK), Polyamides (nylon), Polycarbonate, Polybutylene Terephthalate (PBT) and Polytetrafluoroethylene (Teflon) are the most common engineered thermoplastics. High-performance polymers display special characteristics such as heat, fire, chemicals, UV and other resistance. Polysulfones, polyether-sulfonates, polyphenylene sulphide, liquid crystal polymers, polyimides, etc. are common high-performance thermoplastics. Commodity thermoplastics are produced in bulk and are low cost and low performance compared to engineered and high-performance thermoplastics [11, 13].

Thermoplastics polymers are categorised into the following three groups based on the degree of crystallinity [14]:

1. Crystalline thermoplastics including polypropylene (PP), low-density polyethene (LDPE) and high-density polyethene (HDPE).
2. Amorphous thermoplastics such as polystyrene (PS), polyvinyl chloride (PVC), polymethylmethacrylate (PMMA), polycarbonate (PC) and acrylonitrile-butadiene-styrene (ABS).
3. Semi-crystalline polymers including polyester polybutylene terephthalate (PBT) and polyamide Imide (PAI).

The recycling techniques for thermoplastics can be categorised into primary, secondary, tertiary and quaternary recycling processes [15]. The primary and secondary recycling techniques are based on mechanical techniques while the tertiary and quaternary recycling techniques are based on chemical techniques.

4.1 Primary Recycling Techniques

The primary recycling process is the most common, simple and cost-effective. In this process, plastic products are reused in their original structure to produce new products. The disadvantage is that each material can be reused for a limited number of cycles. The polymeric waste is collected, sorted and washed before cutting, grinding and shredding. The small sized pieces of the polymeric waste are reprocessed into a product with or without mixing with virgin materials. The advantage of this technique is reduced cost of raw materials, however, the post-consumer polymer needs to be stabilised to avoid degradation during the processing.

4.2 Secondary Recycling Techniques

Secondary recycling techniques are also termed mechanical recycling methods. The thermoplastic polymer waste is re-melted and reprocessed into new products without the alteration of polymer structure during mechanical recycling. In this process, the plastic waste is collected, sorted, washed and shredded into granules, flakes or pallets followed by melting to produce new products via the extrusion route. Floatation, gravity, electrostatic or magnetic separation and sensor based sorting methods can be adopted for separation and sorting of collected thermoplastic waste. The recycled materials can also be blended with virgin material for reduced cost and improved properties [16].

The disadvantages of mechanical recycling are the heterogeneity of solid waste, reduction in quantity and the degradation of material properties due to the lower molecular weight of recycled polymer waste. The reduction in molecular weight occurs during the chain-scission reactions due to the presence of moisture and acidic impurities. The lowering of molecular weight can be avoided by extensive washing and drying of waste, compounding with chain extender chemicals and reprocessing with vacuum degassing. The operational cost of secondary recycling processes is lower but initially requires substantial capital investment to establish the process facility.

4.3 Tertiary Recycling Techniques

Tertiary recycling is a set of chemical processes that allows the conversion of polymers into smaller molecular chains and monomers or promote the partial depolymerisation of polymers to oligomers. The resulting monomers can be re-polymerised to produce original or related polymers. In this process, the thermoplastic waste polymer is reduced to molecules that can be used as starting materials such as monomers, oligomers and/or a mixture of hydrocarbon compounds for feedstock [17]. The major chemical reactions that can be enacted to decompose the polymers into monomers are listed in Table 2.

The chemical processing routes for recycling require huge capital investment, expertise and a skilled workforce. Most of the chemical recycling processes are under investigation such as gasification and pyrolysis and are not upscaled to a commercial scale of applications. However, several processes such as glycolysis and methanolysis are currently practised on a commercial scale.

PET (Polyethylene Terephthalate) Polymer can be cleaved into different products by using different reagents and reactions such as hydrolysis, acidolysis, glycolysis or alcoholysis [18]. During hydrolysis, PET is made to react with acidic or alkaline or neutral water for complete depolymerization into monomers. The hydrolysis method is, however, very costly for its high temperature (in the range of 200–250 °C), high pressure (in the range of 1.4–2 MPa) and longer time to complete depolymerisation.

Several approaches are adopted in the past to hydrolyse PET flakes. For alkaline hydrolysis of PET, PVC and PVC coated fabrics NaOH or KOH solutions of 4–20% concentration are used. The autoclave temperature ranges from 120 to 200 °C when aqueous NaOH solutions are used, while 110–120 °C when a non-aqueous solution of KOH in methyl cellosolve is used. 2% admixture of isophthalic acid and 98% pure TPA (terephthalic acid) can be obtained using alkaline hydrolysis at the

Table 2 List of major chemical reactions that can be utilised to decompose the polymers into monomers

Tertiary recyling techniques	Hydrogenation
	Glycolysis
	Gasification
	Hydrolysis
	Pyrolysis
	Methanolysis
	Chemical depolymerisationsition
	Thermal cracking
	Catalytic cracking and reformaing
	Photo degradation
	Ultrasound degradation
	Degradation in microwave reactor

initiation energy of 99 kJ/mol. Sulphuric acid favours the separation of TPA of high purity. Sulfuric acid, nitric acid or phosphoric acid are used for acidic hydrolysis and depolymerisation of PET waste in the temperature range of 70–100 °C. Hot water, steam, or pressurised autoclave (1–4 MPa) at the temperature (in the range of 200–300 °C) are used for neutral hydrolysis. This process yields around 95% TPA if the temperature is above 245 °C while complete depolymerisation takes place at 275 °C. Among many hydrolysis techniques, neutral hydrolysis is more eco-friendly and thus has attracted more interest, however, the mechanical impurities that are present in the polymer are left in TPA. This results in the product with considerably lower purity compared to the products produced with acid or alkaline hydrolysis [19, 20].

Glycolysis is a chemical recycling method in which PET chains are converted into bis(hydroxyethyl) terephthalate (BHET), a common substrate for PET synthesis, by glycol insertion [21]. Ionic and basic ionic catalysts are used during PET glycolysis. 1-butyl-3-methylimidazolium ([Bmim]OH) is a hydroxyl-based basic ionic liquid and exhibits higher catalytic activity for PET glycolysis, compared to bicarbonate ([Bmim]HCO_3), chlorine ([Bmim]Cl) and bromide ([Bmim]Br) based 1-butyl-3-methylimidazolium. Factors including catalysis, reaction temperature and polarity of the reaction mixture influence the reactivity order of different glycols. The purification of the products in the glycolysis process catalysed by ionic liquids is easier and simple compared to the traditional metal acetate catalysts [22, 23].

Methanolysis involves the use of methanol for PET depolymerisation into dimethyl terephthalate (DMT) and ethylene glycol (EG) at a temperature ranging from 180 to 280 °C and pressure ranging from 2 to 4 MPa [24]. The rate of methanolysis is influenced by the extent of PET solubility. When the main products are obtained by methanolysis, the DMT is distilled and reused to produce net PET.

Polyolefin is another major group of thermoplastics that has found applications in toys, electronic components, films, containers, cases, etc. Unlike polymers like PET, they are not recycled into monomers from which they are produced as a result of random scission of the C–C bonds. Olefins are usually recycled through a pyrolysis technique in which the polymeric waste is degraded by heating in an oxygen-free environment. The oil and gaseous fractions of aliphatic composition are retrieved during the pyrolysis of LDPE, PP and HDPE and can be recycled as a feedstock for new polymers in the petrochemical industry [25]. Olefins can also be recycled through mechanical (dissolution/reprecipitation) and chemical recycling (pyrolysis) techniques. The prior approach leads to higher recovery of the pure fraction of polymer but at a cost of many organic solvents which makes the latter approach more promising. The decomposition rate of PP is higher compared to the LDPE and HDPE. The polymers with a lower amount of crystalline region and more branched structures display lower stability in thermal degradation [26].

Supercritical solvents including benzene, toluene or ethylbenzene are used to depolymerise PS polymers into styrene at a temperature ranging from 310 to 370 °C and at 6.0 MPa pressure. However, toluene is more effective than benzene and ethylbenzene if the degradation process is carried out at 360 °C for 20 min [27].

4.4 Quaternary Recycling Techniques

Quaternary recycling techniques are aimed to recover energy from plastic waste. Controlled incineration of plastic waste is a common way to recover energy content from plastics and reduce the volume of waste. The flue gases produced during incineration are very toxic such as dioxins which are not environmental friendly. Sometimes they are used as a co-fuel for heating kilns, but a similar problem persists. Incineration is a simple and quick solution to recover energy from polymer waste if the flue gas is treated with appropriate air pollution control methods before discharge into the environment [28].

Another approach to reusing polymeric waste is to mix it with gravel and bitumen at around 160–170 °C to build plastic roads, which have a longer life than conventional gravel-bitumen roads. Crushed plastic waste is also melted and mixed with sand to produce sand bricks, tuff tiles, etc, however, such methods can lead to an increase in microplastic pollution [29].

Among the several recycling techniques discussed above, the chemical recycling of thermoplastic waste complies with the principles of sustainable development to recover the monomers for polymer feedstock.

4.5 Applications of Recycled Thermoplastics

Recycled plastics are used for several applications including the food industry, electrical and electronic appliances and other engineering structures. The biggest problem in reusing plastic waste for different applications is the possible presence of chemical or microbiological contaminants from the previous use and sometimes from the pre-recycling process such as washing with detergents [30]. The relative regulatory authorities have reported their reservations regarding the use of recycled plastics, particularly for applications in the food industry. The contaminants including the molecules of plastics, used for food packaging or in contact with food, must not migrate or appear in the food product. The recycled plastics may not comply with the regulations for food contact due to the potential health hazards. However, the migration of undesired contaminants from packaging into a food can be avoided by:

1. Appropriate post-use washing if the same packaging is reused in its form.
2. Depolymerisation of plastics to reproduce fresh polymer for packaging from recovered monomers.
3. Using multilayer plastics.

Using virgin polymer for the layer that meets food and using recycled plastic for the outer layer can be an alternative solution. The internal layer plays the role of a functional barrier and avoids the migration of contaminants to food from the recycled plastic layer [31].

The sorption and diffusion properties of plastics vary according to the polymer type and are very important when plastics are selected for the food industry. The polymers that exhibit better inertness such as poly ethylene naphtholate (PEN), PET and rigid PVC are more appropriate for food packaging compared to PS, HDPE, PP and LDPE [32].

Recycled polymers are used extensively for indoor applications. In the recent past, the environmental risks associated with plastics used for home appliances such as air conditioners, heaters, refrigerators, blowers, computers, cell phones and other consumer electronics have increased. The European Union has issued directives regarding plastic waste from electrical and electronic components. The use of several hazardous substances is prohibited while alternative environmental friendly designs are suggested that allow the use of recyclable and energy-efficient polymers. The recycling yield of plastics used in such equipment has increased since these regulations are implemented. The recycled plastics used in the electrical appliance are recovered from high-grade plastics. For example, the back of television body casing, flow fans, balancers and base frame for washing machines, dewatering tanks and protection nets of air conditioners are produced using recycled PP and PS. Recycled PP is also used to produce food cases for refrigerators and water tanks for washing machines. Moreover, they are also used in producing several automotive and aerospace parts for their simple repair, casting and joining possibilities [33].

5 Recycling of Thermosets

The production volume of thermosets is much lower compared to thermoplastics, but they are important materials for high-performance applications in the automotive, aerospace and military industries. Most thermoset polymers are reinforced with a significant number of reinforcements such as glass, carbon, aramid or other fibres. Polyurethane (PU) is the most used thermoset polymer in the industry followed by unsaturated polyester, phenolic resin, amino resin and epoxy [12].

Thermoset polymers are generally considered difficult materials to recycle and reprocess due to the presence of a covalent intermolecular network that is responsible for improved strength and stiffness, creep, thermal and chemical resistance compared to thermoplastic polymers. Thus, thermoset polymers are the most promising materials for application in structural and protective industries such as composites for automotive, aerospace and other critical engineering structures. On the other hand, the higher thermal and chemical resistance makes the thermosets very difficult to recycle [34]. Table 3 shows the major groups of thermoset resins.

The increasing demand for lightweight thermoset materials in engineering structures for their high-performance and energy efficiency along with the increasing regulations on landfills are the driving forces behind the quest to develop more sustainable methods for thermoset recycling.

The direct reuse of recycled plastic materials for applications where a lower quality is acceptable such as housing and sports equipment to avoid the reprocessing steps. Since thermoset materials have non-adoptable geometries, therefore, direct reuse methods of plastic waste do not contribute significantly to solving the problem of thermoset waste. The conventional techniques to recycle thermosets are based on mechanical, thermal and chemical processing [34, 35].

5.1 *Mechanical Recycling Techniques*

Currently, most of the thermoset plastic waste materials ends-up in landfills which is the least preferred way of waste management. The most common methods to reuse thermoset polymers and composite materials waste are grinding and incineration. The grounded thermoset polymer and composite powder are used as a filler in other thermoset polymer composites or rubbers. During the grinding process, the

Table 3 Major groups of thermoset polymers

Thermoset resins	Isocyanates
	Unsaturated polyesters
	Formaldehydes
	Epoxies
	Alkydes

constituents and molecular structure of the thermoset waste are not changed but it, however, downgrades their properties of reinforcements [34, 36].

5.2 Thermal Recycling Techniques

The recovery of reinforcement fibres is an alternate option and can be achieved if the matrix component of the thermoset composite is decomposed thermally or chemically. However, this process can be economically viable if reinforcement material is highly valuable. The recovered carbon fibres from the waste composites can retain around 90% of the original properties. Thermal processing techniques for recycling can be categorised into the following two groups [37]:

1. Aerobic combustion (incineration).
2. Anaerobic combustion (pyrolysis).

The incineration leads to the recovery of only energy and no other useful material is recovered. However, there are several modified methods of incineration proposed in the recent past. Pickering et al. [38] proposed a fluidised-bed aerobic combustion technique for the recovery of reinforcement (glass fibres) from Glass Fibre Reinforced Polymer (GFRP) composites and the thermoset matrix polymer is consumed to recover energy. GFRP scraps can also be used as feedstock for co-processing in cement kilns where the matrix component of the thermoset composites is combusted while the glass fibres, composed of borosilicate and calcium carbonate, are utilised as raw material/feedstock for cement production. This is so far the most economical solution commercially in practice for the recycling of GFR thermosets. The required process energy for pyrolysis has a substantial influence on the viability of thermoset polymer waste recycling.

Aerobic combustion (pyrolysis) breaks down the thermoset polymer into lower molecular weight gaseous, liquid and carbon-based molecules correlating to their monomers. The pyrolysis is carried out at a temperature ranging from 300 to 800 °C. During the pyrolysis, the properties of carbon fibres, used as reinforcement in Carbon Fibre Reinforce Polymer (CFRP) thermoset composites, are not affected due to oxidation and can be reused several times [39].

The properties of carbon fibres in CFRPs are not affected by oxidation. The efficiency of thermoset pyrolysis can be improved with the use of specific catalysts such as fly ash, which itself is a waste of coal power plants and thus makes pyrolysis a more commercially viable thermal recycling process.

5.3 Chemical Recycling Techniques

The recycling techniques in which specific chemicals as used to decompose or dissolve the thermoset polymers are termed Solvolysis. This approach allows the potential recovery of reinforcement material and the constituents materials of the thermoset matrix to be reused as a resource for polymers. Solvolysis techniques can be categorised into hydrolysis, glycolysis and acid digestion based on the applied dissolution medium. Several thermoset polymer linkages including amine-epoxy, anhydride-epoxy, polyester and polyurethane polymers can be cleaved and hence recycled using solvolysis techniques. In the recent development of recycling medium, supercritical fluids have proved to be good solvents for the degradation of thermoset polymers due to their high mass transport coefficients, high diffusivities, low viscosities and no thermal energy requirement and controllable reaction rates [40, 41].

Most of the solvolysis techniques proposed in the past are performed at a temperature higher than 200 °C, therefore, these techniques are more often considered thermo-chemical recycling techniques. The use of hazardous chemicals in these processes is also a downside. The solvolysis approach to recycling thermoset waste is still under investigation and is not commercially in practice [42].

5.4 *Inherent Recyclability in Thermosets*

The approaches to recycling thermoset polymers, mentioned earlier in this chapter, often require a higher amount of energy and/or chemicals with a complex process to recover the resin. An alternative way to recycle thermoset plastics and their composites is to recover the matrix polymer instead of its constituents. Inherent recyclability in thermoset polymers can be achieved by incorporating liable or dynamic covalent bonds into the conventional thermoset polymers and can be further altered by the amount and position of liable bonds in the structure.

The integration of liable bonds within the molecular structure of thermosets is a technique that enables the reuse, reshaping, self-healing or recycling of these resins. The dynamic binding mechanism allows the low-energy molecular debonding and re-cross-linking of the thermosets when subjected to stimuli such as temperature, pressure, chemical or optical for easy matrix removal or recycling [43, 44].

In general, the mechanism for inherent recyclability in thermosets can be categorised into the following three groups:

(a) The complete degradation of thermoset polymer network into lower molecular weight molecules (gaseous, liquid and/or solid compounds such as char).
(b) The degradation of the polymer chain (backbone) to give monomers or oligomers.
(c) The degradation of cross-linking bonds to allow cross-linking such as in thermoplastic polymers.

The stimuli triggered degradation of polymers destroys the polymer architecture and the recyclate (constituents recovered) needs to be resynthesized into polymers. Alternatively, the thermosets with inherited dynamic covalent networks function via associative or dissociative mechanisms that allow the re-cross-linking to produce a cross-linked structure like the original and display mechanical properties almost similar to the parent thermoset polymer [45, 46].

Cleavable crosslinks allow the triggered rearrangement of crosslinks but do not degrade even at high loadings. Shieh et al. [47] incorporated cleavable bonds within the strands of polydicyclopentadiene (pDCPD) thermoset polymer. The resulting polymer displayed properties similar to native material with the ability to undergo triggered degradation. This technique allows the controlled degradation of thermoset polymers. However, the thermosets resin modified with inherent recyclability for degradation or integrated with dynamic covalently cross-linked chemistries are relatively expensive and complex which makes them industrially irrelevant for the moment.

Unlike composites, the interest of researchers in recycling other polymeric products is not satisfactory, for example, rubbers for car tyres, etc. Rubbers are thermosets with relative different properties from resins but due to permanent covalent linkages, they are also difficult to recycle. However, they are not usually reinforced with fibres, therefore, the recycling process can be simpler. Rubber thermoset polymers can be efficiently recycled by mechanical and thermal recycling techniques but this approach is highly energy consuming. Dynamic covalent linkages can be introduced in rubbers via disulfide linkages which are already present due to the vulcanisation process, however, the associative nature of disulfide linkages in rubbers makes the recycling process very complex [48].

The dissociative DA linkages for rubbers are extensively studied and exhibit high possibilities for recycling and reprocessing. Thermal interface materials were produced with easy repairability via associative disulphide linkages. While repairable flexible electronics were developed utilising dynamic DA and polyamine linkages in the recent past. These repair abilities are stimulated by heat, which is already applied in internal use, however, the moisture is avoided as an environmental hazard. With such technologies, the commercial development of products using recycled and/or reprocessed thermosets is possible in near future [49, 50].

5.5 *Applications of Recycled Thermosets*

Unlike thermoplastics, the thermosets market is more fragmented, and it is difficult to find reliable data on the market size. However, based on the reported data, the global production of thermoset polymers is around 65 million tonnes per year, which is around 20% of the global production of polymeric materials [47]. Thermosets are preferred materials to produce lightweight fibre-reinforced composites for high-performance and efficient engineering structures. Glass fibres are the most used reinforcement in thermosets due to their cost-effectiveness. Carbon fibres are relatively expensive and are used for high-performance applications. Due to the growing demand for lightweight materials in energy-efficient engineering structures, the overall growth in thermoset polymer is highly expected in the coming decades.

The increase in demand for thermosets especially in automotive and aerospace industries will ultimately result in an increased amount of thermoset polymer waste and higher demand for recycling [51].

Due to the much lower specific gravity of thermosetting plastic compared to the typical fine and coarse aggregates, thermosets such as melamine plastic waste is used in the non-loadbearing lightweight concrete [52]. Ford's automotive industry reported in June 2018 that each of its vehicles is integrated with recycled plastics and has consumed 250 bottles per vehicle. This way, they are reusing around 1.2 billion bottles per year.

Most of the recycling techniques, that are in practice for thermoset polymers rely on a higher amount of energy and/or chemicals and are focused on the recovery of valuable reinforcement instead the polymer itself. All the approaches to recycle thermoset recycling, discussed above, have their pros and cons and reveal that there is no universal solution to the problem of thermoset waste. Therefore, for the improvement of recyclability, each product is focused on separately. It is important to consider the end application of the recycling product before adopting a specific recycling approach.

The recycling concepts based on degradable or dynamic thermosets, especially for composites, are more application-oriented approaches as more valuable fibres and resins are reclaimed. The composites produced with the recycled components have displayed performance comparable to their commercial counterpart. Sporting goods, automotive parts and wind turbine structures are successfully produced with recycled resins.

GFR composites with industrially acceptable mechanical performance with added healing properties are being produced using thermoset epoxy resins with inherent recycling characters such as trimer, polyamine or disulphide linkages. However, different advantages associated with these techniques are matrix-fibre binding, matrix-fibre separation during recycling in case of vitrimer and disulphide linkages induced polymers [53].

References

1. Umoren, S.A., Solomon, M.M., Saji, V.S.: Chapter 1—basic polymer concepts I. In: Polymeric Materials in Corrosion Inhibition, pp. 3–48. Elsevier
2. Ashfaq, A., Clochard, M.-C., Coqueret, X., et al.: Polymerization reactions and modifications of polymers by ionizing radiation. Polymers (Basel) **12** (2020). https://doi.org/10.3390/polym12122877
3. Reddy, M.S.B., Ponnamma, D., Choudhary, R., Sadasivuni, K.K.: A comparative review of natural and synthetic biopolymer composite scaffolds. Polymers (Basel) **13** (2021). https://doi.org/10.3390/polym13071105
4. Burford, R.: Polymers: a historical perspective. J. Proc. R. Soc. New South Wales **152**, 242–250 (2019)
5. Silviya, E.K., Varma, S., Unnikrishnan, G., Thomas, S.: Compounding and mixing of polymers. In: Thomas, S., Weimin, Y. (eds.) Advances in Polymer Processing, pp. 71–105. Woodhead Publishing (2009)

6. Gad, S.E.: Polymers. In: Wexler, P. (ed.) Encyclopedia of toxicology, 3rd edn., pp. 1045–1050. Academic Press, Oxford, Third Edit (2014)
7. OECD.: Global plastics outlook: plastic waste in 2019. Environ. Stat. (2022). https://doi.org/10.1787/a92f5ea3-en
8. OECD: Global plastics outlook: economic drivers. OECD Publishing, Paris, Environmental Impacts and Policy Options (2022). https://doi.org/10.1787/de747aef-en
9. Cabernard, L., Pfister, S., Oberschelp, C., Hellweg, S.: Growing environmental footprint of plastics driven by coal combustion. Nat. Sustain. **5**, 139–148 (2022). https://doi.org/10.1038/s41893-021-00807-2
10. Fotopoulou, K.N., Karapanagioti, H.K.: Degradation of various plastics in the environment. In: Takada, H., Karapanagioti, H.K. (eds.) Hazardous Chemicals Associated with Plastics in the Marine Environment, pp. 71–92. Springer International Publishing, Cham (2019)
11. Merrington, A.: 9—Recycling of plastics. In: Kutz, M. (ed.) Applied Plastics Engineering Handbook, 2nd edn., pp. 167–189. William Andrew Publishing, Second Edi (2017)
12. Utekar, S., Suriya, V.K., More, N., Rao, A.: Comprehensive study of recycling of thermosetting polymer composites—driving force, challenges and methods. Compos. Part B Eng. **207**, 108596 (2021). https://doi.org/10.1016/j.compositesb.2020.108596
13. Ashraf, M.R.: Commodity polymers, engineering polymers and speciality polymers. AcademiaEdu 1–30 (2015)
14. Wagner, J.R., Mount, E.M., Giles, H.F.: 19—Polymer structure. In: Wagner, J.R., Mount, E.M., Giles, H.F. (eds.) Extrusion, 2nd edn., pp. 225–232. William Andrew Publishing, Oxford, Second Edi (2014)
15. Chohan, J.S., Singh, R.: Thermoplastic polymers: a review on primary, secondary, tertiary and quaternary recycling. In: Hashmi, M.S.J. (ed.) Encyclopedia of Materials: Plastics and Polymers, pp. 558–566. Elsevier, Oxford (2022)
16. Bevis, M.: Secondary recycling of plastics. Mater. Des. **3**, 344–349 (1982). https://doi.org/10.1016/0261-3069(82)90096-6
17. Lee, A., Liew, M.S.: Tertiary recycling of plastics waste: an analysis of feedstock, chemical and biological degradation methods. J. Mater. Cycles Waste Manag. **23**, 32–43 (2021)
18. Stanica-Ezeanu, D., Matei, D.: Natural depolymerization of waste poly(ethylene terephthalate) by neutral hydrolysis in marine water. Sci. Rep. **11**, 1–7 (2021). https://doi.org/10.1038/s41598-021-83659-2
19. Karayannidis, G.P., Achilias, D.S.: Chemical recycling of poly(ethylene terephthalate). Macromol. Mater. Eng. **292**, 128–146 (2007). https://doi.org/10.1002/mame.200600341
20. Grigore, M.E.: Methods of recycling, properties and applications of recycled thermoplastic polymers. Recycling **2**, 1–11 (2017). https://doi.org/10.3390/recycling2040024
21. Guo, Z., Lindqvist, K., de la Motte, H.: An efficient recycling process of glycolysis of PET in the presence of a sustainable nanocatalyst. J. Appl. Polym. Sci. **135**, 6–11 (2018). https://doi.org/10.1002/app.46285
22. Bassi, M.A., Lopez, M.A., Confalone, L., et al.: Chemical recycling of PET wastes with different catalysts. Nature **388**, 539–547 (2020)
23. Maurya, A., Bhattacharya, A., Khare, S.K.: Enzymatic Remediation of Polyethylene Terephthalate (PET)–based polymers for effective management of plastic wastes: an overview. Front. Bioeng. Biotechnol. **8**, 1–13 (2020). https://doi.org/10.3389/fbioe.2020.602325
24. Kurokawa, H., Ohshima, M., Sugiyama, K., Miura, H.: Methanolysis of polyethylene terephthalate (PET) in the presence of aluminium tiisopropoxide catalyst to form dimethyl terephthalate and ethylene glycol. Polym. Degrad. Stab. **79**, 529–533 (2003). https://doi.org/10.1016/S0141-3910(02)00370-1
25. Sogancioglu, M., Yel, E., Ahmetli, G.: Pyrolysis of waste high density polyethylene (HDPE) and low density polyethylene (LDPE) plastics and production of epoxy composites with their pyrolysis chars. J. Clean. Prod. **165**, 369–381 (2017). https://doi.org/10.1016/j.jclepro.2017.07.157
26. Shoaib, M., Subeshan, B., Khan, W.S., Asmatulu, E.: Catalytic pyrolysis of recycled HDPE, LDPE, and pp. Prog. Rubber, Plast. Recycl. Technol. **37**, 264–278 (2021). https://doi.org/10.1177/14777606211019414

27. Ke, H., Li-hua, T., Zi-bin, Z., Cheng-fang, Z.: Reaction mechanism of styrene monomer recovery from waste polystyrene by supercritical solvents. Polym. Degrad. Stab. **89**, 312–316 (2005). https://doi.org/10.1016/j.polymdegradstab.2005.01.014
28. Schwarz, A.E., Ligthart, T.N., Godoi Bizarro, D., et al.: Plastic recycling in a circular economy; determining environmental performance through an LCA matrix model approach. Waste Manag. **121**, 331–342 (2021). https://doi.org/10.1016/j.wasman.2020.12.020
29. Shajahan, S., Rajan, S.P.: A Review Paper on Usability of Different Waste Plastic Types on Bitumious Concrete Mix, pp. 4696–4699 (2019)
30. Oksana, H., Andrea, C., Daniel, D.-L., Andrés, F.: Applications and future of recycling and recycled plastics. In: Parameswaranpillai, J., Mavinkere Rangappa, S., Gulihonnehalli Rajkumar, A., Siengchin, S. (eds.) Recent Developments in Plastic Recycling. Springer Singapore, Singapore, pp. 345–372 (2021)
31. Triantafyllou, V.I., Karamani, A.G., Akrida-Demertzi, K., Demertzis, P.G.: Studies on the usability of recycled PET for food packaging applications. Eur. Food Res. Technol. **215**, 243–248 (2002). https://doi.org/10.1007/s00217-002-0559-1
32. Thorsheim, H.R., Armstrong, D.J.: Recycled plastics for food packaging. Chemtech August 55–58 (1993)
33. Reis, J.P., de Moura, M., Samborski, S.: Thermoplastic composites and their promising applications in joining and repair composites structures: a review. Mater (Basel, Switzerland) **13** (2020). https://doi.org/10.3390/ma13245832
34. Xu, X.-Q., Liao, S., Wang, Y.: Recycling of thermosetting plastics. In: Parameswaranpillai, J., Mavinkere Rangappa, S., Gulihonnehalli Rajkumar, A., Siengchin, S. (eds.) Recent Developments in Plastic Recycling. Springer Singapore, Singapore, pp. 95–119 (2021)
35. Pickering, S.J.: Recycling technologies for thermoset composite materials—current status. Compos. Part A Appl. Sci. Manuf. **37**, 1206–1215 (2005). https://doi.org/10.1016/j.compositesa.2005.05.030
36. Delva, L., Ragaert, K., Kuzmanovic, M., et al.: Mechanical recycling for dummies. Plast. TO Resour. 1–26 (2017)
37. Yansaneh, O.Y., Zein, S.H.: Recent advances on waste plastic thermal pyrolysis: a critical overview. Processes **10** (2022). https://doi.org/10.3390/pr10020332
38. Xue, X., Liu, S.-Y., Zhang, Z.-Y., et al.: A technology review of recycling methods for fiber-reinforced thermosets. J. Reinf. Plast. Compos. **41**, 459–480 (2022). https://doi.org/10.1177/07316844211055208
39. Abdallah, R., Juaidi, A., Savaş, M.A., et al.: A critical review on recycling composite waste using pyrolysis for sustainable development. Energies **14** (2021). https://doi.org/10.3390/en14185748
40. Kuang, X., Zhou, Y., Shi, Q., et al.: Recycling of epoxy thermoset and composites via good solvent assisted and small molecules participated exchange reactions. ACS Sustain. Chem. Eng. **6**, 9189–9197 (2018). https://doi.org/10.1021/acssuschemeng.8b01538
41. Kulikova, Y., Sliusar, N., Korotaev, V., et al.: Recovery and use of recycled carbon fibers from composites based on phenol-formaldehyde resins. Recycling **7**, 22 (2022). https://doi.org/10.3390/recycling7020022
42. Hu, J., Dong, H., Song, S.: Research on recovery mechanism and process of waste thermosetting phenolic resins based on mechanochemical method. Adv. Mater. Sci. Eng. (2020). https://doi.org/10.1155/2020/1384194
43. Bangash, M.K., Ruiz de Luzuriaga, A., Aurrekoetxea, J., et al.: Development and characterisation of dynamic bi-phase (epoxy/PU) composites for enhanced impact resistance. Compos. Part B Eng. **155**, 122–131 (2018). https://doi.org/10.1016/j.compositesb.2018.08.039
44. Urdl, K., Kandelbauer, A., Kern, W., et al.: Self-healing of densely crosslinked thermoset polymers—a critical review. Prog. Org. Coat. **104**, 232–249 (2017). https://doi.org/10.1016/j.porgcoat.2016.11.010
45. Gil, E.S., Hudson, S.M.: Stimuli-reponsive polymers and their bioconjugates. Prog. Polym. Sci. **29**, 1173–1222 (2004). https://doi.org/10.1016/j.progpolymsci.2004.08.003

46. Paipa-Alvarez, H.O., Palacios Alvarado, W., Medina Delgado, B.: Biodegradable thermosets polymers as an alternative solution to pollution generated by plastics. J. Phys. Conf. Ser. **1672** (2020). https://doi.org/10.1088/1742-6596/1672/1/012013
47. Shieh, P., Husted, K.E.L., Johnson, J.A.: Cleavable comonomers enable degradable, recyclable thermoset plastics. Annu. Tech. Conf. ANTEC, Conf. Proc. 542–547 (2021). https://doi.org/10.1038/s41586-020-2495-2.Cleavable
48. Mohd Sani, N.F., Yee, H.J., Othman, N., et al.: Intrinsic self-healing rubber: a review and perspective of material and reinforcement. Polym. Test. **111**, 107598 (2022). https://doi.org/10.1016/j.polymertesting.2022.107598
49. Canadell, J., Goossens, H., Klumperman, B.: Self-healing materials based on disulfide links. Macromolecules **44**, 2536–2541 (2011). https://doi.org/10.1021/ma2001492
50. Xiang, H.P., Rong, M.Z., Zhang, M.Q.: Self-healing, reshaping, and recycling of vulcanized chloroprene rubber: a case study of multitask cyclic utilization of cross-linked polymer. ACS Sustain. Chem. Eng. **4**, 2715–2724 (2016). https://doi.org/10.1021/acssuschemeng.6b00224
51. Post, W., Susa, A., Blaauw, R., et al.: A review on the potential and limitations of recyclable thermosets for structural applications. Polym. Rev. **60**, 359–388 (2020). https://doi.org/10.1080/15583724.2019.1673406
52. Mohammadian, M.: A study on application of recycled thermosetting plastic in concrete. Rev. Rom. Mater. Rom. J. Mater. **43**, 276–284 (2019)
53. Yue, L., Bonab, V.S., Yuan, D., et al.: Vitrimerization: a novel concept to reprocess and recycle thermoset waste via dynamic chemistry. Glob. Challenges (Hoboken, NJ) **3**, 1800076 (2019). https://doi.org/10.1002/gch2.201800076